Geographies of Disorientation

Spatial disorientation is of key relevance to our globalized world, eliciting complex questions about our relationship with technology and the last remaining vestiges of our animal nature. Viewed more broadly, disorientation is a profoundly geographical theme that concerns our relationship with space, places, the body, emotions, and time, as well as being a powerful and frequently recurring metaphor in art, philosophy, and literature.

Using multiple perspectives, lenses, methodological tools, and scales, *Geographies of Disorientation* addresses questions such as: How do we orient ourselves? What are the cognitive and cultural instruments that we use to move through space? Why do we get lost? Two main threads run through the book: getting lost as a practice, explored within a post-phenomenological framework in relation to direct and indirect observation, wayfinding performances, and the various methods and tools used to find our position in space; and disorientation as a metaphor for the contemporary era, used in a broad range of contexts to express the difficulty of finding points of reference in the world we live in.

Drawing on a wide range of literature, *Geographies of Disorientation* is a highly original and intriguing read which will be of interest to scholars of human geography, philosophy, sociology, anthropology, cognitive science, information technology, and the communication sciences.

Marcella Schmidt di Friedberg is Professor of Geography at the 'Riccardo Massa' Department of Human Sciences and Education at the University of Milano–Bicocca, Italy. Her research interests include cultural geography, island studies, gender geography, and the history of geographical thought. She is a member of the Editorial Board of ACME and Chair of the IGU Commission on the History of Geography.

Geographies of Disorientation

Marcella Schmidt di Friedberg

Routledge
Taylor & Francis Group
LONDON AND NEW YORK

First published 2018 by Routledge

2 Park Square, Milton Park, Abingdon, Oxfordshire OX14 4RN
52 Vanderbilt Avenue, New York, NY 10017

Routledge is an imprint of the Taylor & Francis Group, an informa business

First issued in paperback 2019

Routledge is an imprint of the Taylor & Francis Group, an informa business

British Library Cataloguing in Publication Data
A catalogue record for this book is available from the British Library

Library of Congress Cataloging in Publication Data
A catalog record for this book has been requested

ISBN: 978-1-4724-5048-7 (hbk)
ISBN: 978-0-367-24525-2 (pbk)

Typeset in Times New Roman
by Taylor & Francis Books

For Alfi, who never gets lost

Contents

Tables

Acknowledgements

This book is the outcome of many years of wandering through the meanders of a theme that I consider to be of the utmost importance. I am particularly grateful to Claude Raffestin for having encouraged and advised me along the way, and for sharing my passion for the labyrinth. I thank Guido Scarabottolo for granting me permission to reproduce his drawing on the theme of getting lost. Several scholars contributed valuable feedback on draft versions of chapters and chapter sections: amongst others, Claudio Smiraglia, Günther Seufert, Enrico Squarcina, Giuseppe Iaria, and Augusto Lovagnini Scher. The "Riccardo Massa" Human Sciences and Education Department at Milano-Bicocca University provided financial support enabling the timely completion and translation of this work. I am also indebted to Kathy Crossan and the editorial team at Routledge for believing in this project and its publication. I thank Clare O' Sullivan for her scrupulous and thoughtful translation, and my master's and doctoral students in Anthropological and Ethnological Sciences and Social and Cultural Anthropology, respectively, for sharing their insights with me in the course of lengthy discussions on the theme of disorientation. I would also acknowledge the use of extracts of the following publications with grateful thanks: John O'Keefe and Lynn Nadel (1978), *The Hippocampus as a Cognitive Map*, originally published by Oxford University Press, by permission of the authors; Sara Ahmed (2006), *Queer Phenomenology: Orientations, Objects, Others*, by permission of Duke University Press; Mark Monmonnier (1991), *How to Lie with Maps*, by permission of University of Chicago Press; Allen W. Wood, *Kantian Ethics* (2008), by permission of Cambridge University Press; Paul A. Dudchenko (2010), *Why People Get Lost: The Psychology and Neuroscience of Spatial Cognition*, by permission of Oxford University Press; Reginald G. Golledge, ed. (1998), *Wayfinding Behavior: Cognitive Mapping and Other Spatial Processes*, by permission of Johns Hopkins University Press.

I am grateful to my daughters, Caterina and Margherita, for their moral support and for sharing with me the joy of getting lost. Finally, disorientation is such a broad theme that I owe thanks to many more people – colleagues, friends, and students (Elena dell'Agnese, Cristina Schmidt, Marie-Clare Robic, Marjorie Hedges, Silvia Lista, Valeria Pecorelli, Mario Neve, Stefano

Malatesta, Margit Schreiner, Pia Antonini, Karine Douplitzky, Marco Benedetti, Aylie Lonmon, Fabio Castelli, Rossella Giraudi, Giorgio Mosterts, Anna di Seyssel, and numerous others), who – often via a single comment or reading recommendation – opened my eyes to a whole new dimension of disorientation.

Note: The English translation of the text, including citations from the foreign-language sources listed in the bibliography, is by Clare O'Sullivan.

Foreword

Claude Raffestin

Our choices almost invariably conceal an enigma, for ourselves in the first instance and for others thereafter. Why did we choose to do this rather than that? In most cases, the question is of greater import than the answer, because the question is the very foundation on which knowledge is built. Over time, answers may vary, but the questions will remain the same.

Why has a geographer developed an interest in – I might say a passion for – the labyrinth? Claiming that the answer is less important than the question exonerates me from seeking out or imagining all the reasons behind Marcella's decision to grapple with the labyrinth; in reality, however, it was the labyrinth that grabbed hold of her – though happily she has managed to avoid being devoured by the Minotaur!

Oh, how well I understand this passion of Marcella's for the labyrinth, which concentrates and crystallizes the essence, if not the entirety, of all questions of space and time. For what is the labyrinth if not a symbolic microcosm which at once comprehends *both* the material place of a relational system, woven among multiple actors of shifting morphologies whose limits attest to their existence, *and* the immaterial place of a set of great myths that have been incessantly revisited, reworked and reinterpreted throughout history? A symbolism that is amply illustrated in this book, to which no foreword can have anything to add. As so aptly observed by Borges: "a preface, when favoured by the stars, is not in the manner of a toast; it is a lateral form of criticism." Thus, I draw on lateral thinking here in order to throw into relief what might be concealed at first view, rather than to eclipse this volume's evident riches.

Orientation-disorientation, space-place, movement-dance, body-spirit – such are the guiding themes of this book, which fully engages with the myth of the labyrinth, whether natural or constructed, that has obsessed the history of thought and art for centuries and continues to inspire all sorts of literary and artistic creations – works that cast new light on that which we believed to be already established, revealing previously unknown facets. One cannot help thinking of Koch's famous fractal curve, which gives the impression of being finite because it can always be inscribed within a given circumference, but which becomes longer at each successive iteration and therefore has an infinitely long perimeter! Metaphorically speaking, it is legitimate to expect that the myth of the labyrinth can generate an infinite number of situations and interpretations, in which the same starting coordinates are arranged in ever-new configurations.

Truly the labyrinth fascinates, has ever fascinated and will ever continue to fascinate us! It is undoubtedly the mirror that deforms many of the questions to which we are unable to find answers: it is the simplest and perhaps the most effective form of disorientation, and model *par excellence* of all mistaken relations whose inevitable outcome is death. The labyrinth is also deeply paradoxical, in the etymological sense of the term: on the one hand, it is a place that is haunted, indeed inhabited, by the death that is consummated and encountered there; yet, it originated as an architectural structure for hiding away the consequence of an excess of life and energy, in other terms, a passion that needed to be channelled, confined and "disoriented" in order to be mastered. Passion is confinement to the extent that it deprives a being of its capacity to think. It is confinement, and consequently disorientation, when it provokes a loss of reference points and rationality, as Kant so convincingly demonstrated.[1] The problem posed by the labyrinth is that posed by life and death in all their forms.

Not alone a paradox, but clearly also an aporia, the labyrinth is a myth of surprising richness. Is it a place of good and/or evil? It is both, verging on the status of undecidable aporia. Let us recall the sequence: irrefutable passion combining both the purest and least pure forms of desire, in the throes of which Pasiphae allows herself to be taken by the bull, who has nothing to offer except his powerful and instinctive bestiality, having no will to seduce. Pasiphae is the seductress who needs an artifice supplied by Daedalus, himself perennially seduced by any enigma needing to be resolved. Daedalus too is consumed with passion, a passion for solving all problems that come his way. Although his role in the unfolding drama is neither direct nor thrilling, it is Daedalus who holds the key to the action, who facilitates the passage from one state of equilibrium to another. Through his inventions, he never definitively solves anything, but he sets the action in motion once again, coordinating scenes that without him would never have formed the particular concatenation that constitutes the greatness of the myth. Daedalus' inventiveness is stimulated by the need for artifices, which in turn are required to forget and flee from an original sin that clearly cannot be cancelled out. He manages to delay the tragic outcome, but remains imprisoned himself. For no one ever emerges unharmed from the labyrinth, whose form and spirit marks the destiny of each of the actors in the myth! A prison ordered by Minos but constructed by Daedalus, the labyrinth captures the Minotaur, confining him and isolating him both physically and psychologically. This said, we ourselves play the part of all these heroes simultaneously, because in the final analysis, our lives are nothing other than the construction of our own labyrinths! A tragic example is the life of Walter Benjamin, who got out of his own personal "labyrinth" by committing suicide in Portbou.

This leads us to reflect that, when it comes to "sense of direction", there is a genealogy of the labyrinth to be written, which is precisely what Marcella undertakes in this book when she explores the theme of differences between the "savage" and the "civilized" in relation to the navigation of space. Here again, a myth is magnificently invoked – that of Gilgamesh, in which Enkidu "the savage" loses his natural reference points as a result of frequenting the "cultured and civilized" woman. The myth of the labyrinth contains all of this and Marcella has captured it to perfection. In investigating the question of disorientation, she draws attention to the opposition between the "animal part" and the "human part" of culture, or to state this in other – simpler – terms, between the somatic dimension and that of the "logos".[2]

She has been sensitive to the opposition between body and mind, in relation to which, ancient philosophy, following Plato, especially tended to devalue the body. However, the somatic part of the labyrinth is significant and Marcella recognizes this in her elegant evocation of dance. The figures of the dance fluidly trace images of great and chaotic complexity that may be likened to the outline of the labyrinth, given that both dance and labyrinth are the rigorous execution of an arrangement by the choreographer and the architect, respectively. Terpsichore, the "dancer", has no need to blush before her sisters! She might even be proud of her precision in the purest abstraction that she ever inspired or could possibly have inspired… Daedalus' labyrinth! The labyrinth equally evokes the movement of the snake and convolutions of the brain, urban design structures, and initiatory paths in which the body is key, though often forgotten in favour of the religious dimension.

It might be thought that the author has strayed quite far from geography here, or in any case from a certain type of geography, and this is all the better, because if she has moved away, it has been to embrace a fundamental "geographicity" – one that concerns itself with the large-scale migratory movements of recent years, whereby Europe has become the forced labyrinth in which migrants from the East suffer, despair and die. They have been evicted from their homes by nameless wars, and by civil wars initiated and maintained by the irresponsible and unworthy political leaders of peoples who find themselves hostage to selfish and hypocritical interests. No, the Minotaur is not dead and he continues to feed on the sacrificial victims that he finds along the pathways of the European labyrinth. This time, who will play the part of Theseus the liberator?

As we can see, the ancient myths tend to repeat themselves, and at hitherto unprecedented levels of monstrosity. Thus, suffering gives rise to disorientation, which first and foremost entails confusion and a breakdown in the dimensions of space and time. At the individual level, this is identified with the well-known Korsakoff syndrome, yet groups and communities are also affected although their symptoms are generally neither described nor named. There is a delirium, or state of confusion, that is found in individuals, and is well known to psychiatrists, but there is also one that afflicts societies, as reflected in many of the news stories reported daily by the tabloid press.

Marcella's chosen theme could not be of greater, or more painful, relevance than today, and this should put us on our guard and alarm us: the power of the myth is immeasurable! Comenius made no mistake in calling the story of his seventeenth-century tribulations the "Labyrinth of the World". The labyrinth, as viewed and treated by Marcella, fits perfectly with one of Hans Blumenberg's key ideas: "Metaphor is an aesthetic medium, because it resides in the pre-existing sphere of the concept. It is continuously required to compensate for the concept's inadequacy, deficits and limited performance."[3] The aesthetic, as Blumenberg observes, emerges from the mythical and the metaphorical. We must raise this question in light of the tendency, growing stronger by the day, for the concept to lose popularity to the image. The concept tends to pale in the face of images, and therefore, is it not the case that it needs the mythical and the metaphorical to come to its aid and help it recover its powers of persuasion? Is it not the case that we have underestimated the role of aesthetics – in the structural, that is to say, in the ethical, sense of the term – in political discourse?

No doubt, these are a lot of questions, but such is the price to be paid if we are to bring the fundamental to light. For if the labyrinth has retained its fascination for us over thousands of years, this is because it is nothing other than questions with a proliferation of responses that reflect the fears and anxieties of each new era. The contemporary relevance of the labyrinth is also given by its role as a place that generates emotions. The study of emotion is not only fashionable, it has become crucial: emotion is a key component of knowledge inasmuch as the labyrinth contributes to both the world of knowledge and the world of emotions. This ambivalence represents the very richness of the myth, I would be tempted to say, of all myths. Freud's work was in great part inspired by this ambivalence, and no matter what certain of our contemporaries think, I am not sure that Freudian theory has lost value to the extent that is often claimed.

The labyrinth poses questions about its own origins and signification, and Marcella does not fail to touch on this. All the existing explanations and etymologies are skilful and ingenious, but none of them put an end to the mystery or preclude the craziest of conjectures. We are forced to conclude that, similarly to the famous two-headed axe, *Labrys*, the labyrinth always offers both a conceptual side and a metaphorical side. The latter assists the former by endowing it with the poetic substance it needs to survive and evolve. Reading Marcella's book makes one suspect, although she does not say so, that she regrets geography has not yielded more to the metaphoric side in its conceptualizing. Clearly, she does not explicitly engage with this hypothesis, yet a book that did not stimulate the reader to "imagine" could not be entirely successful…

I cannot but wish good fortune to this work, which joins a long tradition of theorizing about the genealogies of the labyrinth, in which Pliny the Elder laid down a milestone, one century after Christ, along a path that leads uninterruptedly to our own day.

Notes

1 Cf. Kant (2001).
2 See Shusterman (2008) for an interesting discussion on the current revaluing of the body.
3 See Blumenberg (2010, p. 32).

Introduction

The Aleph

In *Who Am I: An Autobiography of Emotion, Mind, and Spirit* (2009), the Chinese-American geographer Yi-Fu Tuan recounts that when asked at social gatherings about his reasons for becoming a geographer, he generally replies with some vague statement about his love of travel. While most people are satisfied with this "lazy answer", there is a second, more thoughtful response:

> I have always had an inordinate fear of losing my way. Of course, no one likes to be lost, but my dread of it is excessive. I suspect that more than physical discomfort is at stake. To be lost is to be paralyzed, there being no reason to move one way rather than another. Even back and front cease to be meaningful. Life, with no sense of direction, is drained of purpose. So, even as a child, I concluded that I had to be a geographer so as to ensure that I should never be disoriented. Geographers always know where they are, don't they? They always have a map somewhere – either in their backpack or in their head.
>
> (Tuan, 2009, p. 93)[1]

Dis-orientation is a profoundly geographical theme that concerns our relationship with space, places, the body, emotions, and time. Losing our bearings, if only for an instant, is an experience shared by most human, and non-human, beings. Our ability to orient ourselves draws on a complex set of components from the cognitive-intellectual, emotional, sensory-motor, cultural and social spheres. So, how exactly do we find our way? What are the cognitive and cultural instruments that we use to move through space? Why do we get lost? Disorientation is a broad and transversal theme that speaks to a multiplicity of disciplines ranging – in addition to geography – from neurobiology to the cognitive sciences, anthropology and sociology, city planning and architecture, information technology and the communication sciences. Disorientation is also a powerful and frequently recurring metaphor in art, philosophy, and literature.

My aim here is not to provide a definitive analysis of this vast topic, based on field research or on one or another univocal methodological approach. Rather, I use multiple perspectives, lenses, methodological tools and scales to explore some alternative interpretations of the phenomenon of getting lost: by definition an ambiguous theme that does not fit into any single scientific or disciplinary framework. From the physical and neurological perspective, many questions about the orientation of humans and non-humans remain unanswered:

> Despite the evidence of a direct relationship between increased neural activity in specific brain structures and the use of selective spatial information for orientation, to date, the neural mechanisms responsible for the complex behavioural phenomenon of orientation remain unclear. One reasonable explanation [...] is the complexity of this cognitive process.
>
> (Arnold et al., 2013, p. 360)

Within geography, reflection on orientation/disorientation cannot but start from the notion of space and from the geographical subject, a subject that is becoming increasingly more fragmented and multiple. Space, or rather "the practice and knowledge [...] of the phenomenon that we refer to as space" (Raffestin, 1981, pp. 268–269), is at the heart of geographical reflection as its most characteristic component, while no discourse about orientation and disorientation is possible without making reference to space. From the multiple perspectives of the new cultural geography, space is no longer a mere container, nor "the dead, the fixed, the non-dialectical, the immobile" (Foucault, 2001, p. 70), but is open and interacts with subjects and objects.

> Space can never be that completed simultaneity in which all interconnections have been established, and in which everywhere is already linked with everywhere else. A space, then, which is neither a container for always-already constituted identities nor a completed closure of holism. This is a space of loose ends and missing links. For the future to be open, space must be open too.
>
> (Massey, 2005, p. 12)

Disorientation is a powerful metaphor for the contemporary era and is used in a broad range of contexts to express the difficulty of finding points of reference in the world we live in.[2] In the words of French philosopher Bernard Stiegler, author of *Technics and Time, 2: Disorientation* (1996):

> Seen as originary, disorientation is always constituted by identifiable, characteristic orientation-markers (cardinalité) designating its borders, indicating North and South, Orient and Occident. In disorientation, however, Orient and Occident are not simply geographic givens; Orient and Occident designate particular experiences of disorientation. Such

> cardinal directional markers, insofar as they open (to) the horizon of meaning and configure the motif for all motivation, can be reified only through experience of and in the world. […] disorientation opens a space of difference, between here and there, public and private, profane and sacred, strange and familiar, and so on.
>
> (Stiegler, 2009, p. 3)

In Stiegler's view, disorientation is linked to the technical era that began with the Industrial Revolution, and to the speed and pace at which it has expanded, which in turn has given rise to a growing gap between technological development and social organization: "Contemporary disorientation is the experience of an incapacity to achieve epochal redoubling. It is linked to speed, to the industrialization of memory resulting from the struggle for speed, and to the specifics of the technologies employed in that struggle" (ibid., p. 7). The feeling of disorientation and the need to negotiate a new belonging are also part of the experience of migration: "Migration could be described as a process of disorientation and reorientation: as bodies 'move away' as well as 'arrive', as they reinhabit spaces" (Ahmed, 2006, p. 9).

In the technical era, orientation is delegated to a machine – the GPS, "a prosthesis"[3] that enables us to identify our exact position at all times and to plan itineraries without getting lost; however, this "increased presence" of technology and the corresponding excess of information can themselves lead to disorientation:

> How can we really live if there is no more here and if everything is now? How can we survive the instantaneous telescoping of a reality that has now become ubiquitous, breaking up into two orders of time, each as real as the other: that of presence here and now, and that of a telepresence at a distance, beyond the horizon of tangible appearances?
>
> (Virilio, 1997, p. 37)

In relation to the contemporary world and its complexity, Lefebvre invokes an empty space, a contradictory space, characterized by: "fragmentary contents, […] incoherence under the banner of coherence; a cohesion grounded in scission and disjointedness, fluctuation and the ephemeral masquerading as stability" (Lefebvre, 1991, pp. 308–309); our orientation is thus constructed around the question, both material and metaphorical, "Where am I?" and around its multiple and heterogeneous answers. This issue and its implications in today's world bring us back to the fundamental theme of geographical space and to the spatial turn in the social sciences: "And yet, if everything is moving where is here? […] If there are no fixed points then where is here?" (Massey, 2005, pp. 138–139) asks Doreen Massey.

The simultaneity and speed of postmodern spatial experience bring to mind the "infinitesimal" and "impossible" "listing of an endless series" at the core of Borges' tale of the Àleph. In this short story from "the fantasy genre", the final piece in a collection of the same title (1945), Borges imagines having accidentally discovered the limitless Aleph in an old cellar in Buenos Aires: "The only place on earth where all places are – seen from every angle, each standing clear, without confusion or blending" (Borges, 1971, Vol. I, p. 894), giving the following account of the encounter:

> I shut my eyes – I opened them. Then I saw the Aleph [...] In that single gigantic instant I saw millions of acts both delightful and awful; not one of them occupied the same point in space, without overlapping or transparency. What my eyes beheld was simultaneous [...] On the back part of the step, towards the right, I saw a small iridescent sphere of almost unbearable brilliance [...]. The Aleph's diameter was probably little more than an inch, but all space was there, actual and undiminished. Each thing (a mirror's face, let us say) was infinite things, since I distinctly saw it from every angle of the Universe.
>
> (ibid., pp. 10–14)

In the dazzling infinity of the Aleph, Borges can discern an endless multitude of things, which are on different scales but viewable simultaneously. Examples of these wondrous sights include the teeming sea, daybreak and nightfall, the multitudes of America, a cobweb in the centre of a pyramid, the splintered labyrinth of London, countless eyes, all the mirrors in the world, equatorial deserts and every last grain of sand in them, horses, tigers, pistons, bison, tides, armies, and all the ants on Earth. Borges concludes:

> I saw the Aleph from every point and angle, and in the Aleph I saw the Earth and in the Earth the Aleph, and in the Aleph the earth; [...] and I felt dizzy and wept, for my eyes had seen that secret and conjectured object whose name is common to all men but which no man has looked upon – the unimaginable universe.
>
> (ibid., pp. 10–14)

The "unimaginable universe" of Borges' Aleph is the starting point for Edward Soja's reflection on the "Trialectics of Spatiality", prompted in turn by Lefebvre's notion of the "spatial triad": "One always has Three. There is always the Other" (Lefebvre, 1980, p. 143). In Soja's view, *The Aleph* is an invitation to adventure as well as to modesty, but above all is: "an allegory on the infinite complexity of space and time" (Soja, 1996, p. 57). Inspired by Borges' imaginings, Soja introduces the concept of Thirdspace: "The space where all places are, capable of being seen from every angle, each standing clear; but also a secret and conjectured object, filled with illusions and allusions, a space that is common to all of us yet never able to be completely seen and understood, an 'unimaginable universe'" (ibid., p. 56). As in a deck of cards, in Thirdspace our usual reference categories and their political and ideological frameworks are shuffled to form a "dizzying array of different kinds of spaces" (ibid., p. 59). Soja continues: "Everything comes together in Thirdspace: subjectivity and objectivity, the abstract and the concrete, the real and the imagined, the knowable and the unimaginable, the repetitive and the differential, structure and agency, mind and body, consciousness and unconscious, the disciplined and the transdisciplinary, everyday life and unending history" (1996, pp. 56–57).

How does one move around and find one's bearings in third space? "Trajectoral thinking is difficult, for it challenges all conventional modes of thought and taken-for-granted epistemologies. It is disorderly, unruly, constantly evolving, unfixed, never presentable in permanent constructions" (ibid., p. 70): in other words, it invites us to embrace disorientation. In a space that is "equally exhilarating and threatening", getting lost acquires new meanings and may become, as in the Aleph, an experience of terror or joy, signifying either existential anguish or a space of freedom. Let us now more systematically examine the possible interpretations of getting lost and the implications – both pleasant and unpleasant – of disorientation.[4]

The "uncanny feeling" of getting lost

According to Californian writer[5] Rebecca Solnit, author of *A Field Guide to Getting Lost* (2005), the word "lost" comes from the Norse *los*, which means: "the disbanding of an army, and this origin suggests soldiers falling out of formation to go home, a truce with the wide world" (Solnit, 2005, p. 7). The dark and troubling image of soldiers without a guide, bewildered by a momentary interruption in the hell that is war, evokes the different senses in which one may be disoriented. Getting lost is an act that is rich in meaning, predominantly understood as having a negative valence, or as a condition of extreme anxiety. As Kevin Lynch describes it:

> To become completely lost is perhaps a rather rare experience for most people in the modern city. We are supported by the presence of others and by special way-finding devices: maps, street numbers, route signs, bus placards. But let the mishap of disorientation once occur, and the sense of anxiety and even terror that accompanies it reveals to us how closely it is linked to our sense of balance and well-being. The very word "lost" in our language means much more than simple geographical uncertainty; it carries overtones of utter disaster.
>
> (Lynch, 1960, p. 4)

The first meaning of getting lost is therefore a negative one. The terrifying experience of going beyond the confines of the known world, of losing sight of our usual points of reference also features in the "analysis of cultural apocalypses" offered by Italian philosopher and anthropologist Ernesto de Martino: "Two antithetical terrors inform the age we live in: that of 'losing the world' and that of 'being lost in the world'" (De Martino, 1977, p. 475).

The "bell-tower of Marcellinara" has become the symbol of this form of existential anguish and sense of disorientation. De Martino, in *La fine del mondo*, recounts an experience of his own in 1950s' Calabria, when he and a friend stopped in the village of Marcellinara to enquire for directions from an elderly shepherd. To preclude any possibility of error, De Martino asked the shepherd to accompany them as far as the turning that they needed to take a few kilometres ahead:

> He got into the car somewhat reluctantly, as though he feared some kind of trap, and this reluctance gradually grew into distress, because from the car window, out of which he was constantly peering, he had lost sight of the bell-tower of Marcellinara, the point of reference for his extremely circumscribed domestic space. Due to the disappearance of the bell-tower, the poor old man felt completely disoriented: and only with difficulty did we manage to get him to the right turning and to obtain the information that we needed.
>
> (ibid., 1977, pp. 480–481)

At last they brought the shepherd back to the village. By now he was in a state of great anxiety, and, in his impatience to see the bell-tower to reappear, travelled with his head leaning out of the car window: "When he finally saw it, his face relaxed and his aged heart began to quieten, as though he had won back a 'lost homeland'" (ibid., 1977, pp. 480–481).

On the party's return to Marcellinara, the old man jumped out of the car without so much as a goodbye, deeply disturbed by the unpleasant adventure that had rudely snatched him from his "existential space", as identified by the familiar image of the Marcellinara bell-tower. Existential anguish due to the loss of familiar everyday points of reference that are also linked to tradition – the significance of the bells as a religious symbol[6] is not to be overlooked – may also occur in relation to place in the modern era. According to De Martino, a similar anguish of displacement may be suffered by astronauts during space travel:

> When in the cosmic silence they lose their connection with the "bell-tower of Marcellinara" that is planet Earth, and the world of men: and they speak and speak without pause [...] to avoid losing their "sense of belonging to the earth". Surely our presence becomes at risk when it touches the limit of its existential homeland, when it loses its "bell-tower of Marcellinara".
>
> (ibid.)

And indeed, such a sense of disorientation has featured in the history of lunar exploration. The famous explosion aboard Apollo 13, caused by an electrical fault in the oxygen tank, disrupted the functioning of the craft's computer navigation system. The astronauts were obliged to fall back on using a sextant to work out their position by observing the stars. However, a shower of debris from the explosion made it impossible for them to distinguish stars from particles glimmering in the sunlight. At this point, they had to rely on NASA's backup navigation plan and their only remaining means of getting back to Earth, which involved using the Earth's terminator, or the line that separates the portion of the Earth where the sun is shining from the portion in darkness, as a guide to position. During the moon landing of the crew that manned Apollo 14, the astronauts experienced, for the first time, the phenomenon of Spatial Disorientation (SD), defined as the "failure to correctly perceive attitude, position, and motion", due to difficulty in adjusting to the lack of atmosphere and the unfamiliar inclination of the horizon (Clark et al., 2010, p. 5).

The "tragedy" of disorientation described by Yi-Fu Tuan, Lynch, and De Martino may be linked to one of the categories used by Freud to analyze, as Derrida has documented (*To Speculate - on "Freud"*, 1985, pp. 236–258), the ambivalence of pleasure and the notion of "negative pleasure". In a 1919 essay *Das Unheimliche* (*The Uncanny*[7]) that pre-empts the later *Beyond the Pleasure Principle* (1921), Freud examines the wholly subjective aesthetic experience of feelings of "unpleasantness and repulsion". The uncanny, for Freud

> is a province of this kind. It is undoubtedly related to what is frightening – to what arouses dread and horror; equally certainly, too, the word is not always used in a clearly definable sense.
>
> (Freud, 1919, p. 219)

From a linguistic viewpoint, the term "uncanny", chosen by the English translators, does not fully convey the meaning of the original word in German, which contrasts a positive, domestic, and familiar state (*heimlich*) with the frightening and sinister state of the *Un-heimliche*. To Freud, what is "uncanny" is frightening precisely because it departs from what is known and familiar: "The uncanny is that class of the frightening which leads back to what is known of old and long familiar" (ibid., p. 220). Its link with the unusual endows the term with a contradictory and ambiguous meaning:

> So that the uncanny would always, as it were, be something one does not know one's way about in. The better orientated in his environment a person is, the less readily will he get the impression of something uncanny in regard to the objects and events in it.
>
> (ibid., p. 221)

Here we find a first connection with our theme of interest, reinforced later in Freud's work when he again uses disorientation to explain the concept of uncanniness. On a hot summer's afternoon in 1919, Freud found himself wandering about a disreputable area of a "provincial Italian city". After some time, he realized that he had unintentionally returned to the spot he had initially started out from, where he was now beginning to draw the attention of the locals. A renewed attempt to leave the area again resulted in an involuntary retracing of his steps: "Now, however, a feeling overcame me which I can only describe as uncanny, and I was glad enough to find myself back at the piazza I had left a short while before" (ibid., p. 237). According to Freud, the uncanny also features in other spatial experiences of disorientation or of getting lost in different places. Feelings of helplessness and uncanniness can be elicited by a wide variety of circumstances, which may be physical in origin, such as getting lost in conditions of poor visibility or on wooded terrain, to the extent that: "every attempt to find the marked or familiar path may bring one back again and again to one and the same spot, which one can identify by some particular landmark" (ibid.). Freud also alludes to the visual experience of becoming disoriented in a dark room, which in a different context prompted Kant's reflection on *What does it mean to orient oneself in thinking* (1786) (see Chapter 3, § 3).

Kant's stumbling about in the darkness of an unknown room is taken up, in lighter terms, by Freud in his analysis of the uncanny: "Or one may wander about in a dark, strange room, looking for the door or the electric switch, and collide time after time with the same piece of furniture" (ibid.), an experience, Freud reminds us, that Mark Twain managed to portray as highly comical.[8] The comicality of blundering about a darkened room leads us to the second interpretation of the experience of getting lost, to its positive aspect, or the fact that getting lost may also be *amusing*[9] in the etymological sense of the word:

> amuse (v.) late 15c., "to divert the attention, beguile, delude," from Middle French amuser "divert, cause to muse."
>
> muse (v.) "to reflect, to be absorbed in thought," mid-14c., from Old French muser (12c.) "to ponder, dream, wonder; loiter, waste time," literally "to stand with one's nose in the air".
>
> bemuse (v.) "to make utterly confused."

The delicious flavour of getting lost

The experience of getting lost involves abandoning our familiar points of reference to set out towards the new, and the unknown: "Lost has two disparate meaning. Losing things is about the familiar falling away, getting lost is about the unfamiliar appearing" (Solnit, 2005, p. 22). From being an "uncanny" experience and a "real tragedy", getting lost may alternatively be loaded with expectations, the desire for change, and the longing for novel encounters and explorations; it may become, so to speak, a vocation. As expressed by Doreen Massey: "What space gives us is simultaneous heterogeneity; it holds out the possibility of surprise; it is the condition of the social in the widest sense, and the delight and the challenge of that" (Massey, 2005, p. 106).

The French-Swiss designer Ruedi Baur, who holds trilingual courses and seminars on the theme of *Orientierung/Desorientierung*, proposes a new meaning, which is not necessarily negative, of the term "disorientation": "It must be said that it would certainly have been easier, as is usual in this positivist world, to consider disorientation merely as a disaster [...] even finding more radical solutions to eliminate the blight conclusively and systematically" (2010, p. 12). Baur however proposes disorientation as a tool for breaking with traditional perspectives that bears many positive values:

> Wandering around at random, almost purposely losing track, travelling around, trying to forget where you come from, where you are and where you are heading. [...] Moving somewhere else not from obligation but for pleasure, forgetting your roots, changing your language, profession, family [...] Doing things with no particular purpose, choosing to venture into unknown lands or cultures [...] add one complexity to another, blur distinctions, delve into chaos not to bring in any order but to enjoy getting beyond simplistic models.
>
> (2010, pp. 20–21)

Thus, Baur invites us to learn to get lost, thereby establishing a new kind of relationship with space, in which disorientation becomes a source of potential instead of a limitation. He and his research group are engaged in the search for new directions for dis-orientation, using a range of visual and textual instruments to provide a "Heterogeneous reading of the phenomenon and its positive and negative aspects in progress" (Baur, 2009, Intro.). The Italian anthropologist Franco La Cecla also addresses the theme of getting lost, approaching it from an anthropological and philosophical perspective with an emphasis on the notions of living and localism:[10] "'Getting lost' stands for the episodic or chronic distraction that affects our relations with the environment surrounding us" (La Cecla, 2000, p. 3). In a critical essay on contemporary spatial experience and on the sense of disorientation in the globalized world, La Cecla claims that: "The less we engage with our surroundings the less we are able to orient ourselves in them" (ibid.). Getting lost becomes a practice of resistance to the uniformization of landscapes, to abandonment of the daily ritual of getting to know, producing and organizing places, and to detachment from the practice of living. For La Cecla, getting lost takes two forms: on the one hand, it may be the disorientation of city-dwellers who no longer identify with the familiar image of their city, "a getting lost that is a permanent sense of distraction with respect to one's own territorial context" (ibid., p. 37); on the other, there is a getting lost that is endowed with positive valences, imagination and a sense of discovery, such as the wanderings of the Situationists: "in which getting lost is a practice aimed at restoring value to the undecidable and radically anarchical aspect of spatial experience" (ibid., p. 132). However, this may lead disorientation to become a fashion, a "passion", "a form that is entirely new and pertinent to our times" (ibid., p. 27); getting lost in this sense can even become a commercial offering – introduced into tourist circuits and promoted by travel agencies: "in subordination to laws of the economy of illusion" (ibid., p. 27), whose market value tends to appreciate on account of its scarcity.

As we have seen, disorientation, of its very nature, demands an interdisciplinary approach, insofar as it situates our bodies in relation to the space surrounding us, and the physical, environmental, and cultural relationships between the body and space are mediated by cognitive, psychological, emotional, cultural, and social variables. The theme may be addressed using an enormous variety of methods, underlying philosophical assumptions, scales of reference, and time dimensions. We have seen how voices from different disciplines and scholarship traditions all point up the contemporary relevance of the topic, given its transversal nature, which lends itself to developing an understanding of our relationship with space and the changes currently impacting on our forms of living. Reflecting on disorientation leads us to: "accept complexity, embrace detours, and engage with singularity, to approach them as riches and not as obstacles [...] in short to approach disorientation as a true object of enquiry" (Baur, 2010, p. 12).

The positive value of getting lost is incrementally highlighted by Solnit, Baur, and La Cecla, – a writer, a designer, and an anthropologist – each in his or her own way. In the words of Solnit: "The question is how to get lost. Never to get lost is not to live, not to know how to get lost brings you to destruction, and somewhere in the terra incognita in between lies a life of discovery" (Solnit, 2005, p. 14). For Ruedi Baur the experience of disorientation – a "passion" for La Cecla as we have seen – becomes a "privilege", or an opportunity to savour the "delicious flavour of getting lost": "Under conditions of peace and liberty playing with the loss of orientation can enlarge the horizon of experience and joy" (Baur, 2009, p. 4).

Let us return to the original meaning of the term "lost" or the dismay of the soldiers in disarray. The same image of chaotic desperation in the wake of battle, we are told by Massimo Recalcati, is to be found in the etymology of the word "desire"[11]. In Julius Caesar's *De Bello Gallico*, the *de-siderantes* are once again the dispersed soldiers waiting under the night stars for their companions to return from the battlefield:

> More precisely the etymology of the word "desire" derives from gazing up at the stars in the sky while waiting and seeking to find one's way. Sidera is the Latin for stars, while the privative particle de stands for the impossibility of following the route indicated by the stars, and therefore a condition of disorientation, loss of reference points, homesickness and separation, but also a positive warning of a lack of basic necessities, and the state of waiting for and seeking out one's guiding star.
>
> (Recalcati, 2012, p. 17, fn. 3)

Structure of the book and methodology

As we have seen, disorientation is a multidisciplinary theme that draws on a wide range of materials. To eschew the risk of becoming disoriented amidst this astonishing array of physical and conceptual cues, I have chosen to organize the book according to strict theoretical and conceptual criteria. It is thus divided into three sections, each of which explores a set of logically interconnected topics. My aim throughout is to present the state of the art in research on disorientation as well as to stimulate possible new lines of enquiry for this highly complex field.

The first section of the book, "Orientation/Disorientation", introduces disorientation by situating it within the theoretical frames of geography and philosophy. In the first chapter, I focus on disorientation as practice in physical space, discussing direct and indirect observation, wayfinding performance and the different methods and technological tools that we use to find our position in space. In the second and third chapters, I outline theoretical/methodological approaches to the geographical and cognitive study of orientation/disorientation, and review the history of disorientation as a philosophical dilemma, from Kant to phenomenology and post-phenomenology. In geography, the reflection on orientation and disorientation cannot but start from the notion of space: "Space is, in many ways, the key topic of geography. [It] is a historical concept, determined by advances in science and debates in philosophy" (Elden, 2009, p. 262). According to David Harvey, Kant himself defined geography as: "A synoptic discipline synthesizing findings of other sciences through the concept of Raum [area or space]" (Harvey, 2009, p. 32). Indeed it was Kant, writing as a geographer-philosopher, who formulated the earliest reflections on getting lost in space, using the examples of the darkened room and the right and left hands. The exploration of (dis)orientation in this volume is informed by phenomenological and post-phenomenological geographies. The methodological procedure discussed is therefore the description and classification of the various conditions (or *narrative frames* in the terms of cognitive science) under which the "human-as-flesh body" may be categorized in relation to the world in terms of phenomena of disorientation. Consequently, dis-orientation and re-orientation are also examined in relation to migratory journeys and the feeling of being out of place.

From a (post)phenomenological perspective, orientation and disorientation are crucial issues that have to do with our living experience, bodies (Husserl's *Leib* and Merleau-Ponty's *chair*) and emotions. Within such a theoretical framework, the second section of the book, "Lost Subjects", brings into question, in relation to disorientation, the very subject of geographical discourse, a subject who was originally monolithic, white, male, and Western, but now increasingly appears to be open, differentiated, and fragmented. I focus here (Chapters 4, 5, and 6) on three different subjects of spatial disorientation, categorized in terms of gender, white supremacy, and physical ability, respectively, analyzing how our "corporeal schemas" are genderized, racialized, and normalized.

In the third section of the book, "The Labyrinth of the World", devoted to the metaphor of the labyrinth, I bring "place" to the forefront of my enquiry. Indeed for Tim Cresswell: "Place lies at the center of geography's interests. In a commonsense way geography is about places" (Cresswell, 2009, p. 169). The four chapters in this section (Chapters 7, 8, 9, and 10) deploy the labyrinth metaphor to explore four "places" traditionally associated with disorientation: labyrinths as physical places and images, an enigmatic concept for centuries and still today an esoteric symbol for the complex construction of the world (Chapter 7); the city, literary place of the *flâneur*, and mosaic of different and fragmented spaces, in which it is still possible to get lost (Chapter 8); wilderness – forests and seas – as a landscape of loss, mystery, and fear and, at the same time, of initiation (Chapter 9); the virtual world and cyberspace in which, as we lose ourselves in an ocean of data, technologies challenge traditional notions of human, non-human, natural and artificial and, finally, disorientation in the world of art (Chapter 10).

Notes

1 For Tuan, there is also a third and even deeper answer to the question: "Why are you a geographer?" "I took up geography because I have always wondered, perhaps to a neurotic degree, about the meaning of existence" (Tuan, 2009, p. 93).
2 Cfr. Saunders (2016).
3 See Thrift (2008).
4 The theme of disorientation and its multiple interpretations is strongly present in contemporary social studies. See, for example: "Disorientation as spatial experience: the feeling 'of utter disaster' and the 'delicious flavor' of getting lost", session organized by Marcella Schmidt di Friedberg, Fourth International and Interdisciplinary Conference on Emotional Geographies 1–3 July 2013, University of Groningen, The Netherlands, or Martin and Rosello (2016).
5 Disorientation is a recurrent theme in the literature of all eras. Ondrea Ackerman writes that: "Wandering in literature is a familiar story, from Homer's delayed homecoming to Dante's allegorical pilgrimage, from Spenser's chivalric quests and Cervantes' quixotic ones to Wordsworth's 'internalized' journeys across turn-of-the-century European landscapes. It is not until the twentieth century, however, that 'getting lost' becomes an intentional, textual practice of writers such as James Joyce, Gertrude Stein, Ezra Pound, Samuel Beckett, William Carlos Williams, and contemporary poet Robert Grenier. [...] 'Getting Lost' redefines the project of modernism as an assemblage of mobile views that requires a fundamentally new kind of attention – an ongoing exchange between orientation and disorientation that precipitates new ways of seeing, of moving, and of reading" (Ackerman, 2009, pp. 1–2).
6 "It was always to the steeple that one must return, always it which dominated everything else, summing up the houses with an unexpected pinnacle" (Proust, 2006, pp. 80–81).
7 First published in *Imago*, Bd. V, 1919; reprinted in *Sammlung*.
8 See Twain (1880, p. 107).
9 http://www.etymonline.com, *Online Etymology Dictionary* 2001–2013, Douglas Harper.
10 La Cecla uses the term "fare mente locale": an Italian expression normally used in the figurative sense of "concentrating one's thoughts on a particular topic in order

to bring to mind its precise characteristics" but literally meaning "bringing the mind to a particular place" (see the entry *mente* in www.treccani.it).

11 **desire (v.)** early 13c., from Old French *desirrer* (12c.) "wish, desire, long for," from Latin *desiderare* "long for, wish for; demand, expect," original sense perhaps "await what the stars will bring," from the phrase *de sidere* "from the stars," from *sidus* (genitive *sideris*) "heavenly body, star, constellation" from the phrase *de sidere* "from the stars," from *sidus* (genitive *sideris*) "heavenly body, star, constellation" (Online Etymology Dictionary 2001–2013 Douglas Harper, http://www.etymonline.com).

Part I

Orientation/disorientation

Dis-orientation, as we have seen, concerns our relationship with space, places, the body, emotions, and time: it is a broad and transversal field of enquiry that speaks to a multiplicity of disciplines at both the theoretical and practical levels. Losing one's bearings, if only for an instant, is an experience shared by most human and non-human beings. Our ability to orient ourselves draws on a complex set of components from the cognitive-intellectual, emotional, sensory-motor, cultural, and social spheres. How do we orient ourselves? What are the cognitive and cultural instruments that we use to move through space? Why do we get lost? The three chapters in this section introduce the theme of disorientation from a geographical perspective. In Chapter 1 ("Orientation/disorientation in physical space"), I examine disorientation from a practical-physical point of view, exploring the issue of orientation/disorientation in relation to language, culture, choosing suitable points of reference (stars, geographical cues, environmental factors) and instruments for organizing our movement through space, as well as wayfinding strategies. I bring a sociocultural approach to bear on a review of mythological, historical, linguistic, cognitive, and physical perspectives on spatial coordinates and their construction in different cultural contexts (the history of the compass, the wind rose, *feng shui*, the map). After dealing with physical spatial reality, I turn to the theme of orientation/disorientation in relation to thought, situating it within the theoretical reach of geography, and approaching it from a cognitive perspective. Thus, in Chapter 2 ("(Dis)orienting oneself in thinking"), I review wayfinding and way-losing behaviours and the construction of cognitive maps. In Chapter 3 ("Philosophy of disorientation"), I draw on the substantial body of literature that has dealt with the ancient philosophical discussion on space, outlining key theoretical/methodological frameworks for orientation/disorientation, proposed by philosophers from Kant to the phenomenologists and post-phenomenologists. In conclusion, I discuss disorientation and reorientation in relation to the theme of migration.

1 Orientation/disorientation in physical space

Lost in translation

Approaching the topic of disorientation necessarily involves reflecting on orientation and its different meanings, a familiar theme within geographical enquiry. Getting lost invariably represents a radical discontinuity that affects how we perceive our overall being in the world and our habitual system of reference. "What does it means to be oriented?" – asks Sara Ahmed – "How is it that we find our way in a world that acquires new shapes, depending on which way we turn? If we know where we are, when we turn this way or that, then we are oriented. We have our bearings" (Ahmed, 2006, p. 543). The word orientation is related to the present participle of the Latin verb *orior* – "to be born", "rise", or "come up" – and therefore, to the apparent movement of the rising sun. Observation of the sun and stars with a view to organizing space based on a systematic conceptual order is a practice that is common to many cultures but does not necessarily represent an undifferentiated value or universal constant. The ancient quest to identify and represent a cohesive system of spatial reference points is underpinned both by human beings' practical need to navigate their environment and by a vision of the world that is culturally situated and rooted in people's perceptions of their surroundings, of social and political life, of myth and history. In the words of Yi-Fu Tuan:

> The earth's surface is highly varied. Even a casual acquaintance with its physical geography and teeming life forms tells us as much. But the ways in which people perceive and evaluate that surface are far more varied. No two persons see the same reality. [...] No two social groups make precisely the same evaluation of the environment. The scientific view itself is culture bound – one possible perspective among many.
>
> (Tuan, 1974, p. 5)

Three-dimensional vision and laterality are, in Tuan's view, the characteristics that allow human beings to order perceived reality, organizing it into different categories by dividing it into classes or binary opposites (north-south, right-left, land-sea, mountain-valley, male-female, we-you). Language is one of the fundamental categories that we use to define our relationship with space: "Language structure is the product of our interactions with the world around us. The way we build discourses and develop linguistic categories can immediately be derived from the way we experience our environment and use that experience in species-specific communication" (Heine, 1997, p. 3). Language can often express disorientation. For Wittgenstein: "Language is a labyrinth of paths. You approach from one side and know your way about; you approach the same place from another side and you no longer know your way about" (Wittgenstein, 1953, p. 82). In *The Grammar of Space* (1993), Soteria Svorou explores the relations between language and how we position ourselves in the world in terms of cognitive and social categories compared across 26 randomly selected languages:

> Scientific knowledge about the world does not necessarily enter our belief system, and if it does, it requires the passage of centuries. [...] Our belief system rather is formed by our everyday experiences with the physical world and our cultural environment. [...] We describe the sun as "rising" and "setting" with respect to a stable horizon, and the stars as "coming out" at night.
>
> (Svorou, 1993, p. 1)

The conceptualization of space can differ greatly at the cultural level. Bernd Heine, in a cross-linguistic study of the grammar of language, examines – among other themes – the linguistic structure of spatial orientation, identifying four different forms. Deictic orientation, in which indications of places and objects are immediately identifiable by the subject who is speaking or listening (up, down, in front of, behind, left and right); object-deictic orientation in which the deictic centre may be an inanimate object (in front of the cathedral); landmark orientation (up the hill, down the valley) and cardinal orientation, which is defined in absolute terms, independently of who is speaking or listening (north, south, east, west). Heine adds: "Now, while many cultures do in fact distinguish between deictic and cardinal orientation, some are claimed to lack such a distinction – that is, these cultures are said to have no deictic orientation and/or no terminology for it" (Heine, 1997, p. 14). The human body, as first suggested by Kant (see Chapter 3) is the primary reference point for deictic orientation, followed, in some cases by reference to animal bodies (zoomorphic orientation). Laterality also plays a key role, and the distinction between right and left is never neutral. For Kant, the left side is "weak and worthless", while the right side "enjoys an indisputable advantage over the other in respect of skill and perhaps strength too" (Kant, 1992, p. 392). In his classical study on "religious polarity" the anthropologist Robert Herz adds:

> To the right hand go honours, flattering designations, prerogatives: it acts, orders, and takes. The left hand, on the contrary, is despised and reduced

> to the role of a humble auxiliary: by itself it can do nothing; it helps, it supports, it holds. The right hand is the symbol and model of all aristocracy, the left hand of all common people.
>
> (Hertz, 1960, p. 335)

Alice Werner's research on over 300 Bantu languages (Werner, 1904) demonstrated that the right hand was considered superior (male, strong, big) and the left inferior: "the notion 'left' is frequently etymologically opaque. Wherever a conceptual source can be reconstructed, however, that source is either 'female hand' or else an expression denoting some entity or quality judged to be of inferior status" (Heine, 1997, pp. 48–49). Use of the left hand is connected with right-hemisphere dominance in the brain and vice versa. "Inferior beings", such as women and racial others, are supposed to be associated with the left hemisphere of the brain: "This lack of neutrality is what grounds the distinction between right and left: the right becomes the straight line, the left becomes the origin of deviation" (Ahmed, 2006, p. 14). In addition to the body, place also provides geographical-environmental points of reference, which may be used dynamically depending on how we act or move in space. According to the theory of so-called linguistic relativity, which is based on the observations of Sapir and Whorf but had been anticipated some 100 years beforehand by Wilhelm von Humboldt, language influences our way of conceptualizing the world and our very thinking:

> We dissect nature along lines laid down by our native languages. The categories and types that we isolate from the world of phenomena we do not find there because they stare every observer in the face; on the contrary, the world is presented in a kaleidoscopic flux of impressions which has to be organized by our minds – and this means largely by the linguistic systems in our minds.
>
> (Whorf, 1956, pp. 212–213)

Spatial terms help us to organize our perceptions of our surroundings and supply the bases for our orientation system. However, not all languages feature terms of spatial orientation or organize their vision of space in terms of the cardinal directions. Whorf's classic essay on language and thought among the Hopi of Arizona showed that this culture had no spatial imaginary, nor did it use spatial metaphors (e.g., our positions are distant from one another, a lofty thought, one's mind is wandering): "[u]se of space terms when there is no space involved is NOT THERE – as if on it had been laid the taboo teetotal!" (ibid., p. 146, capital letters in the original). Whorf concluded that the concept of space was apt to vary as a function of language, given its close relationship with other categories of thought, such as time and matter, which are similarly influenced by language usage. In his own words:

> The CONCEPT OF SPACE will vary somewhat with language, because, as an intellectual tool, it is so closely linked with the concomitant

> employment of other intellectual tools [...]. We see things with our eyes in the same space forms as the Hopi, but our idea of space has also the property of acting as a surrogate of non-spatial relationships like time, intensity, tendency, and as a void to be filled with imagined formless items, one of which may even be called "space". Space as sensed by the Hopi would not be connected mentally with such surrogates, but would be comparatively "pure", unmixed with extraneous notions.
>
> (ibid., pp. 158–159, capital letters in the original)

North, South, East, West, the four dwarfs of the Edda

The study of the cardinal points and the familiar scheme of reference that they provide, encompasses cultural and political interpretations, geographical elements, local perceptions and mythological connections, which frequently have little to do with reassuring scientific precision. As expressed by Elisée Reclus:

> On the surface of this round earth the cardinal points have no precise meaning except in relation to particular places. The Greenwich observer may point to his north and his south, his east and his west; but the astronomers of Paris, of Washington, of Santiago, and direction-seeking mankind generally, will look for theirs in other directions. The lines traced by the meridians and the equator are purely artificial. Nevertheless, the attempt has been made to give to the geographical terms of orientation a common meaning that should be accepted by all.
>
> (Reclus, 1894, p. 66)

The main cardinal directions may vary in number across cultures (four, five, eight) or not feature at all. In a study of 127 different languages, Cecil Brown (1983) set out to investigate the origins of the terms used to define the cardinal points, concluding that these origins are not particularly ancient and indeed that: "On a worldwide basis lexical encoding of cardinal directions is a fairly recent phenomenon" (Brown, 1983, p. 122). This claim was based on his observations of the use of spatial terms in traditional societies and how these terms had evolved over time: "The general increase in societal scale and complexity over the last several millennia of human history has no doubt promoted the lexical encoding of cardinal points in many of the world languages" (ibid.). In Brown's view therefore, it is not surprising that more ancient cultures were lacking in terms to indicate the cardinal points, given that these societies were traditionally non-migratory and therefore tended to base their reference systems on local landmarks:

> Technological advances accompanying increases in societal scale such as ocean-going vessels, the compass, maps, mathematics, and so on,

> obviously have contributed significantly to this development. [...] In some instances, cardinal point terms have been directly borrowed from Western languages. In others, native terms have been extended to borrowed concepts of cardinal directionality.
>
> (ibid., p. 123)

Brown identified four main sources of lexicon for describing the cardinal directions: celestial bodies and events, atmospheric phenomena, other more generic directions, and landmarks. Of the 127 languages examined, 81 (about 64 per cent) featured terms for identifying all four cardinal points, while 18 had no such terms. Cardinal point terms were frequently found to be polysemic, that is to say:

> They have referents in addition to cardinal points which seem to be semantically related to them. For example, in addition to left and right, cardinal direction terms often denote other more general directions such as up, down, in front of, behind, and so on.
>
> (ibid., p. 124)

Alternatively, they could refer to celestial bodies or atmospheric phenomena, such as winds, temperatures, weather conditions, and seasons. Brown's analysis showed that north was frequently associated with an upwards direction (in seven cases out of 27 associations), only denoting a downwards direction in one case; vice versa, south was linked to a downwards direction in 11 cases, and only in one instance to an upwards direction (ibid., Table 9, p. 135): "This finding may be indicative of an innate human predisposition for such associations but it is equally plausible that it reflects diffusion of a Western prejudice, i.e. the ubiquitous aligning of north with the top of the maps" (ibid.). The sun, in relation to its rising and setting, was the reference point most commonly shared by the languages under study. The east was most frequently associated with the direction "in front" and the west with the direction "behind". Of 127 languages, 59 (over 45 per cent) connected the east with the rising sun, while 50 related the west to the setting sun (ibid., Tables 2–3, pp. 126–130).

In Latin, as in the majority of languages, orientation was largely based on the apparent movement of the sun, orient-levant and occident-ponente. Again from the perspective of the Ancient Romans, midday and midnight united the temporal with the spatial dimension. The north, on its part, was indicated by the group of seven stars forming the constellation of the Great Bear, for the Ancient Greeks *arktos* (bear) from which the term "Arctic" is derived, for the Latin world *septem triones*, the seven work-oxen that draw the Plough. In oriental Hindu and Buddhist cultures, orientation is under the protection of eight or nine – depending on the specific tradition – powerful divinities, the Guardians of the Directions. The current denomination of the four cardinal points – NSEW – a paradigm of order and precision from the Western

Table 1.1 The main origins of cardinal point terms as surveyed across 127 languages.

	Cardinal Directions					
Origin	North	South	East	West	Total	Percentage
Sun	1	13	58	59	131	57.2%
Deictic orientation	12	13	12	9	46	20.1%
Winds	17	4			21	9.2%
Landmarks	7	10	2	2	21	9.2%
Others	6	3	1		10	4.3%
Total	43	43	73	70	229	100%

Source: Brown (1983), adapted.

cartographic perspective, is derived from Old High German (*Althochdeutsch*) and stems from the creation myth of Norse mythology. In this account of creation, there were nine worlds, forming a diversified and chaotic structure, in an interweaving of temporal dimensions, genealogies, planes and overlapping directions. The system was dominated by the great tree Yggrdrasill, among whose roots, which linked the various worlds, nested the great snake Níðhöggr. Within this complex construction, according to *The Prose Edda* of Snorri Sturluson (dated to around 1220), the cardinal directions took the names of dwarfs, born of the dismembered body of the ice giant Ymir. The Edda recites: "They also took his skull and from it made the sky. They raised it over the earth and under each of the four corners they placed a dwarf. These are called East, West, North and South" (Sturluson, 2005, *Gylfaginning*, Chapter 8, p. 16). Since the thirteenth century, Austri, Vestri, Norðri, Suðri, the four dwarfs of the Edda have indicated the direction of the four cardinal points to the Western world.

The cardinal points take on metageographical valence when they are used to interpret the organization of space in terms of cultural, social, or political values. The division of the world into "North" and "South" introduced by Willy Brandt in 1980[1] reflects a political vision centred around the notion of development, which assigns "rich" countries including Australia, Japan, and New Zealand to the "North" and "poor" countries to the "South", organizing these groups as internally undifferentiated spaces characterized by either wellbeing or global poverty. Equally culturally situated is the contrast between "East" and "West", which paradoxically tends to overlap with the North/South contrast, in that it uses the same binary logic (rich/poor, rational/irrational, organized/disorganized):

> The most general, and in many ways the most influential and pervasive, image or construct we have is that of ourselves and our culture, an entity we have dubbed "Western", a clearly comparative title. Whether it is spoken or unspoken, named or unnamed, we are governed by the notion

> that there is a distinctive cultural history that can be characterized as Western, and it is in distinctive, necessary, and fundamental opposition to non-Western culture and history.
>
> (Menocal, 1987, pp. 1–2)

Historically, the worldview of particular cultures has often been characterized by the choice of a dominant axis of spatial orientation (north/south, east/west). Lindow explores how spatial orientation was conceptualized in ancient Scandinavia, finding that the north–south axis enjoyed privileged status there:

> North was a conceptual category, part of a structure of doubled oppositions, created by culture and bearing meanings imposed on it by culture. The four parts of this structure were necessary for a complete world, a horizontal world with two conceptual axes crossing at the centre, in a structure described by historians of religion as typical of many cultures. [...] in medieval Scandinavia the two axes were not equal: the north-south axis appears to have been primary, and movement along it was charged with primordial and historical (Christian) ethical meaning.
>
> (Lindow, 1994, pp. 220–221)

Lindow argues that the reasons for this preference were to be sought in the geopolitical context of the era: "This is simple geography. The north of Europe was unsettled, the south was populous, and the Church was centered there" (ibid., p. 215). In contrast, the conceptualization of space in early Japan was dominated by an east–west axis, which – according to Palmer – comprehended on the one hand a strong Chinese influence, and on the other, an "indigenous" vision of space related to sun worship (Palmer, 1991).

The position of Europe, and particularly the Greenwich meridian, at the centre of the planisphere has allowed the "Western" world to establish a centre of the world around which to construct logics of power and spatial categories. This Eurocentric vision is evident in Mercator's map projection and the planispheres hanging on our classroom walls. In Mackinder's *Britain and the British Seas*, Great Britain is placed at the centre of the map, in a dominant position from which "the five historic parts of the world are accessible from its waters" (Mackinder, 1902, pp. 3–4). Alternative representations, such as that of Peters, or a different orientation of the map, such as the Australian view with North positioned at the bottom, are disconcerting to the student. We are so accustomed to north-oriented maps that using a differently oriented one leads us to make mistakes, with the map becoming an instrument of disorientation. Thus, the cardinal points may be viewed as symbolic representations that are independent of the geographical reality for which they stand:

> Schoolchildren, irrespective of whether they live in the Amazon basin or the Himalayas, learn that [...] cardinal concepts like "west", "east", "north", and "south" are not expressed in terms of macro-orientation [...]

> but rather in terms of a micro-model introduced in connection with the spread of western formal education.
>
> (Heine, 1997, p. 54)

According to this logic, Asia is all that is not Europe: "It is defined negatively and from the European point of view. [...] The Arabs, the Indians, the Chinese, and the Balinese did not know they were all Asians until the Europeans told them" (Tuan, 1974, p. 42). The terms Levant and Orient are loaded with ideological, political, and social implications:

> The Levant is an area, a dialogue and a quest [...] a synonym for the lands where the sun is rising [...] A Western name for an Eastern area, the Levant was also, by implication, a dialogue between East and West [...] The Levant was also a mentality. It put deals before ideals.
>
> (Mansel, 2010, pp. 1–3)

In a well-known essay, Edward Said argued that the Orient is a European invention: "Since antiquity a place of romance, exotic beings, haunting memories and landscapes, remarkable experiences" (Said, 1978, p. 3). Orientalism as conceptualized by Said is "a manner of regularized (or Orientalized) writing, vision, and study, dominated by imperatives, perspectives, and ideological biases ostensibly suited to the Orient"; this provides a framework for those practices and constructions of thought through which: "The Orient is taught, researched, administered, and pronounced upon in certain discrete ways" (ibid., p. 202). In Said's terms, the *Orient* does not correspond to a real place but to a "system of representations framed by political forces that brought the Orient into Western learning, Western consciousness, and later, Western empire" (ibid., pp. 202–203). The Orient is constructed as the opposite of the West, once again within a binary system of hierarchical relations: "The Orient exists for the West, and it is constructed by the West in relation to the West as a mirror image of what is inferior and alien, or 'Other,' to the West" (ibid., p. 209).

Of stars, winds, and mountains

Orientation is not only a concept and a theoretical formulation of our positioning in the world; it also caters for our practical need to find our way, and to construct a network of familiar reference points in the world around us. The organization of a spatial reference system may be based, as we have seen, on near directions, such as bodies of water, morphological features or significant elements of the landscape (mountains, hills, settlements), and local weather patterns, or on faraway directions, such as the sun and other celestial bodies, the cardinal points and dominant climatic features (prevailing winds, tides, seasonal events), as well as on the values attributed to such factors (good, bad). The choice of preferred reference points – whether elements of the immediate context or cardinal points – is invariably cultural in nature:

"The use of such directions for absolute orientation represents only one possibility; other systems exist in various communities" (Svorou, 1993, p. 30). Language, type of society (nomadic or non-migratory), the morphology of the local territory, technical and scientific knowledge, political and social organization, are all variables that influence which dominant image of the world and system of reference will be chosen. In Tuan's view, this image – which plays a key role in Chinese culture, for example – serves to place the most diverse aspects of our living in the world in relation to one another, organizing them into a connected whole: "In different parts of the world we find the substances or elements, commonly four to six in number, identified with spatial directions, colours, animals, human institutions, and personality traits" (Tuan, 1974, p. 18).

Near directions, drawn from a schematic representation of local morphology, provide a basic orientation system at the local level that is of use to non-migratory populations or human groups whose migratory movements are extremely limited. The link between cultural features and the choice of spatial reference points is emphasized by Tuan, who by way of illustration describes the key role of the Nile in determining the Ancient Egyptians' conception of the world and on the Egyptian sense of direction: "The word 'to go north' meant also 'to go downstream', and the word 'to go south' meant 'to go upstream' or against the current" (Tuan, 1974, p. 86). In this system, the main direction – equivalent to the term "facing", or "in front" – was the south, the source of the great river, while east corresponded to left and west to right. According to Tuan, this led the Egyptians to experience a sense of disorientation in relation to rivers such as the Euphrates that flowed in the opposite direction to the Nile, that is to say, from north to south, as reflected in the highly circuitous manner in which they described these water courses: "That circling water which goes downstream in going upstream" (ibid.).

A further example of the importance of local physical conformation is offered by Heine in describing the image of space held by the Inuit peoples of Greenland who do not use cardinal points as a reference system, but directions related to local geographical elements and prevailing winds: "Since it is likely for rivers to flow and winds to blow in opposite directions on the two sides of Greenland, this means that what is north for a western Eskimo is south for an easterner, and vice versa" (Heine, 1997, p. 51). This system may hinder both orientation itself and communication about direction, even between peoples who are not geographically distant from one another. In a further comment on orientation among the Greenlandic Inuit, Fortescue states:

> Suffice it here to suggest that in certain geographical situations it would be highly confusing for speakers from either side of a strait or deep fjord regularly communicating with each other to have to switch the meaning of directional terms along one or more axis every time they crossed the water, and in others it may be more natural not to change them all.
>
> (1988, p. 6)

Shared images of places recur in geographical and linguistic stereotypes such as the use of the terms "up" and "down" to indicate not merely north and south but also an economic and political hierarchy of place.[2] Our perceptions of the environment surrounding us inevitably include a "social" dimension (Sonnenfeld, 2002) that can vary over time. Joseph Sonnenfeld has studied disorientation in local communities in the south-east and north of Arctic Alaska, a complex environment characterized by extreme weather conditions and long periods of darkness and snowfall. Disorientation varies as a function of how the subject perceives environmental conditions: "Disorientation in any environment can be unnerving. When it occurs in the wilderness, it is either accepted as part of the challenge of a wilderness experience or as more fearsome for the isolation and dangers implicit in the term *wilderness.* In south-eastern Alaska, both natives and non-natives may have difficulties in keeping track of space, time and place, but they appear to respond" (Sonnenfeld, 1994, p. 383). The direction of the winds is another key reference point for the construction of a system of reference. The Bible frequently associates the cardinal points with the winds, four in all, without intermediate directions, each held at a different side of the earth by a guardian angel:

> The north wind is icy cold and wet; from the east comes the warm wind of the Syrian desert [...]; the south wind is warm too [...]; from the west blows a refreshing breeze that carries rain. The winds are messengers and servants of the Lord and storms are [...] instruments of vengeance.
>
> (www.jewishencyclopedia.com)

In Ancient Greek literature (Homer; Hippocrates) two main winds were cited, which later became four (Eurus, Notus, Zephyrus, Boreas) or six; Aristotle, in his *Meteorologia* and Pliny the Elder in *Naturalis historia*, developed an even more complex system of 12 winds. With the expansion of migratory movements and navigation, the winds came to be associated with the cardinal points (Brown, 1983; Aczel, 2001). The poetic image of the wind "rose", formed by 16 or 32 directional rhombi placed around a graduated circle representing the horizon and overlapping like the petals of a rose, began to appear in portolan charts in the fourteenth century, coinciding with the introduction of the compass for navigational purposes; the Catalan Atlas of 1375 provides us with a detailed representation of it. For some, the centre of the rose was the island of Malta, for others the Greek island of Zakynthos. Matvejevic is of the view that the introduction of the wind rose into cartography may have been the work of the Amalfitans, Neapolitans, Arabs or Normans; others attribute its invention to the Mallorcan theologian Raimund Llull. Matvejevic writes that:

> the wind rose is often considered an emblem, but it is more than that. It is spoken of as a metaphor, but it is not only that. [...]. There has been much discussion of sign and symbol in wind roses, their forms and

> meanings, the astrological and calendrical notes accompanying them, the division of the circle into four, eight, sixteen, and thirty-two sections or into twelve and twenty-four, which is temporal as well as geographic [...] The coloured parts of the wind rose are called rhombi on account of their form. In Italy, as the master cartographer Roberto Almagià points out, rhombi with the initials of the principal winds create a circle called a rosone, the word for rosette or rose window, the circular window with roselike tracery found in many cathedrals. [...] The petals of the rose outside the circle may be depicted as torches called feu de joie (fire of joy). [...] There is no reason to doubt that the wind rose originated on the Mediterranean.
>
> (1999, pp. 135–136)

Losing one's bearings: the origin of the compass

Founding myths, political systems, magic, material elements, monuments, colours, psychological characteristics, astral influences, and forms of territory all contribute to shaping our image of space.[3] Tuan compares different cosmological visions (Chinese, Indonesian, Keresan Pueblo, Oglala Sioux), identifying relationships between cardinal directions and colours. In Javanese and Balinese cosmology, depth is added to horizontal spatial orientation to yield a tripartite system: the ethereal world, symbolizing life, is linked to heights, mountains, and dawn; the underground world, a place of sickness, calamity, and death, is associated with the sea and sunset; the intermediate human sphere is located between the other two worlds and is subject to their influence:

> The center is *madiapa*, the intermediate sphere of man which is buffeted by "winds" from both sources. Here, then, in Bali the duality of mountain and sea is mediated by the middle human sphere to constitute a tripartite division; and on the horizontal plane the tripartite division becomes the five-fold schema of four cardinal points and center.
>
> (Tuan, 1974, p. 23)

Hence, nature and society in Java and Bali are structured according to a five-fold schema. For the Balinese, orientation is a crucial cultural construct: "The words for the cardinal points are among the first that a child learns and are used even in the geography of the body. A Balinese will tell you that there is a fly on the 'west' side of your face" (Bateson and Mead, 1942, p. 6). Thus, as observed by Clifford Geertz, in Bali, losing one's orientation is seen as a sign of mental imbalance:

> Paling is social vertigo, the dizzy, disoriented, lost, turned-around feeling one gets when one's place in the coordinates of social space is not clear,

> and it is a tremendously disfavored, immensely anxiety-producing state. Balinese regard the exact maintenance of spatial orientation ("not to know where north is" is to be crazy), balance, decorum, status relationships, and so forth, as fundamental to ordered life (krama) and paling, the sort of whirling confusion of position the scrambling cocks exemplify as its profoundest enemy and contradiction.
>
> (1973, pp. 446–447)

In the vision of the universe developed by Chinese culture over thousands of years, geography, geomancy, religion, the art of numbers, astrology, astronomy, the martial arts, psychology, and architecture lay the foundations for a complex cosmology and a vitalistic conception of nature:

> In a cosmological schema, a substance immediately suggests a color, which suggests in turn a direction, the animal emblem of that direction, and perhaps a human personality trait or mood. In such a richly symbolical world, objects and events take on meanings that to an outsider may seem arbitrary. [...] to the Chinese, "wood", "spring", "east", and "green" implicate each other. The meanings of most symbols are culture-bound.
>
> (Tuan, 1974, p. 23)

Within this system, the ancient geomantic art of Feng Shui, literally "wind and water", is used to harmonically orient roads, buildings, tombs[4] and dwellings, in line with the flows of energy emanating from the earth. At one time the prerogative of a restricted élite of masters (exclusively male), who were expert in religion, geomancy, and other sciences, Feng Shui was declared illegal by the People's Republic of China in 1949, its teachers persecuted and the ancient texts destroyed. Thanks to patient efforts to reconstruct and diffuse what had survived of the ancient tradition, the art of Feng Shui, similarly to other oriental disciplines, is today greatly in vogue in the West, especially in the United States; architecture, design, and furnishings follow its principles in order to create environments in equilibrium with the space around them, as well as in harmony with energies viewed as positive and with inner wellbeing.

According to the Protestant missionary Ernest J. Eitel, active in China during the second half of the nineteenth century and among the first to make Feng Shui known in the West, the Chinese philosopher:

> observes the swift rotatory course of the five planets, Jupiter, Mars, Venus, Mercury and Saturn, and sees their counterpart on earth in the ceaseless interchange and permutations of the five elements of nature, wood, fire, metal, water and earth. He contemplates the spangled firmament at night, and compares with it its dimly-reflected transcript on the surface of our earth, where the mountain peaks form the stars, the rivers and oceans answer to the milky way.
>
> (2012, p. 9)

In Taoist cosmology, bodily organs, colours, seasons, cardinal points and social relations take their place alongside the planets and the natural elements within the complex eight-trigram system of the I-Ching. Taken together, these elements form the Ba-gua (or Pa kua), a symbolic eight-sided figure traversed by the flow of energy from the earth (Ch'i). In Feng Shui the trigrams correspond to geographic and geomantic elements, as well as to the cardinal directions indicated by five celestial animals (the blue dragon to the east, the white tiger to the west, the red phoenix to the south, the black turtle to the north, and the yellow snake in the centre).

The use of the compass – the instrument of orientation par excellence – for divining and magical purposes, predates its use in navigation, from the Mediterranean to China (Aczel, 2002). In Feng Shui, measuring flows of energy (Ch'i) and the balance between Ying and Yang is associated with the Lo p'an, a geomantic compass indicating the south, the most important and propitious direction for Chinese geomancy, and placed at the top of spatial representations. It is not certain whether the invention of the compass for astronomical and geomantic purposes is to be attributed to the Chinese or the Olmec: "The Olmec may have discovered and used the geomagnetic lodestone compass earlier than 1000 BC – predating the Chinese discovery by more than a millennium" (Carlson, 1975, p. 753). Archaeological finds from tombs and other Egyptian, Sumerian, Chinese, and Mesoamerican sites from 4000 BC onwards suggest that the magnetic properties of certain minerals were known from ancient times. Fernand Braudel posed the question: "The compass is supposed to have arrived in the Mediterranean from China in the twelfth century. But is this certain?" (Braudel, 1966, 1, p. 107, n. 20). In the Chinese world, the discovery of the compass is associated with an episode of disorientation experienced by the mythical Yellow Emperor, Huang-Di (2600–2500 BC), one of the founding fathers of Chinese civilization. During the epic battle of Zhuolu against the monstrous Chi You, the legendary yellow Emperor and his army got lost among forests and marshes, on account of the foul-smelling fog breathed out by the demons of the marshes and the mountains, who were allied with Chi You. During the night, the "south-pointing chariot" appeared to the ruler in a dream, enabling him to find his way again and ultimately to win the battle. Apart from the "south-pointing chariot", with its complex mechanical workings, the most ancient directional instrument recorded in China is the Si n'an, a piece of lodestone carved into the shape of a spoon; south-pointing, it was placed on a bronze plate inscribed with the twenty-four directions. Other Chinese directional instruments include directional fish made of metal and immersed in bowls of water, which are mentioned in the *Collection of the Most Important Military Techniques* (*Wujing Zongyao*), dated 1044.

Pointing north or south, as we have seen, the magnetic compass continues to be a valuable and simple instrument for orienting people and things in space for a variety of purposes (Lane, 1963).[5] Its functioning, today enhanced by new technologies, is based on the ability of a needle to align itself with the

Earth's magnetic field, compensating for any deviations. In China, the writings of Gui Guzi and Han Fei (280–233 BC) attest an ancient tradition of studying the orientation of magnetic stones towards the poles. In Europe, the use of the magnetic needle for navigating in bad weather conditions was probably introduced by the Arabs (De Saussure, 1923). The first written mention of it appears in Guyot de Provins (1190), who lamented the fact that the Pope did not fulfil the same role for Christians as the compass needle for sailors; we owe further allusions to the compass to the bishop Jacques de Vitry (1204) and to Brunetto Latini who referred to it as an instrument of long-established use. The analogy between the compass needle and religious or moral teaching recurs in many old literary sources, from the Augustinian monk Alexander Neckham, to Guido Guinizelli, to Dante: "Si mosse voce, che l'ago a la stella parer mi fece in volgermi al suo dove" (Dante, Paradiso, Canto XII, 29–30).[6] The forerunner of modern experimental science in the field of magnetism was Pierre Pelerin de Maricourt, author of the *Epistola de magnete* (1269). The *Epistola* was later followed by the treatise *De magnete* (1600) by William Gilbert, physician to Elizabeth I and the first to introduce the notion of the earth as a huge magnet (Du Trémolet de Lacheisserie et al., 2003). The study of terrestrial magnetism gained new theories and instruments thanks to the work of Carl Friedrich Gauss and Wilhelm Weber. In 1834, with the assistance of Alexander von Humboldt and his numerous international contacts – from the British Royal Society to the tsar of Russia – Gauss and Weber founded the *Göttinger Magnetischer Verein*, a worldwide network of observatories, reaching from Europa to Asia (Cawood, 1979). Using the data from this global grid made up of 84 magnetic observation points, in 1835 Gauss successfully proved Gilbert's hypothesis that the terrestrial magnetic field originates inside the earth itself (Meloni, 1993).

Where is North?

The study of terrestrial magnetism brings us face to face with the highly intricate dynamics of movements both internal and external to our planet and with the difficulty of explaining orientation from a physical point of view. While from a cultural perspective, the invention of the compass may be seen to border on the mythical and magical, from a scientific viewpoint, explaining orientation on the surface of the earth means engaging with the complex physical and magnetic dynamics of the terrestrial system:

> The magnetic poles do not coincide with the geographic poles, they are not antipodal, and their position is subject to irregular variations over time, although these movements are limited and quite gradual, with the poles remaining within the Arctic and Antarctic zones: in 1997 the coordinates of the magnetic north pole were approximately 79° N, 106° W and those of the south pole approximately 64° S, 138° E".[[7]] In 2010, estimates based on the current World Magnetic Model, showed the

> coordinates of the magnetic north pole to be 84.97°N and 132.35°W, and those of the magnetic south pole to be 64.42°S and 137.34°E, with the magnetic north continuing to move about 55 km a year in a north-north-westerly direction.
>
> (The National Geophysical Data Center NGDC, National Oceanic & Atmospheric Administration, NOAA)[8]

This account supposes the interaction of complex geophysical, astronomic, and magnetic processes that are difficult for non-experts to understand, and this may heighten our sense of disorientation. The most up-to-date research on terrestrial magnetism supports a highly complex theoretical framework that is even less accessible to the layperson. As summed up by the International Geographical Glossary: "The lines representing the terrestrial magnetic force field converge at the magnetic poles, which do not coincide with, and are not as stable as, the geographical poles" (Meynen, 1985, p. 162).

Where is the North Pole?[9] Polar orientation – although only the north pole is discussed here, the same is true of the south pole – is based on celestial and terrestrial coordinates deduced – for a variety of purposes – from the movements of our planet. The term "North Pole" may be used to refer, according to the object of our enquiry, to a diverse range of phenomena (Casselman, 2008). The geographic north pole, or true north, is the point in the Arctic Ocean at which the imaginary lines of longitude converge; the difficulty of measuring the precise position of the "true north" is reflected in the long history of attempts to reach it. The airship Norge, flown by the Italian Umberto Nobile, accompanied by Roald Amundsen and Lincoln Ellsworth, flew over the pole on 12 May 1926. It is still a matter of dispute whether the honour of the "discovery" by land should be attributed to the American Frederick Cook (1908) or to his compatriots and fellow explorers Robert Edwin Peary and Mathew Henson (1909), both contenders having been escorted to the region by Inuit on dog-drawn sledges. The first "official" recognition that a party had reached the pole came much later, in 1948, when a Soviet expedition led by Aleksandr Kuznetsov landed nearby and made it to the pole on foot.

The celestial north pole is the point at which Earth's rotational axis, prolonged to infinity, intersects with the celestial sphere. While the other stars visible from the northern hemisphere appear to rotate around this point, the brightest star near the celestial pole appears to be immobile, thus assisting navigation and allowing us to determine at what latitude we are located. The celestial north pole is not fixed; it currently coincides with the North Star at the end of the Little Bear (α Ursae Minoris) constellation, but in the past was represented by Thuban (α Draconis) and still earlier by Vega (α Lyrae); it is expected that Vega will once again indicate the north in about 12,000 years' time: "The Earth's rotational axis does not maintain the same fixed orientation over time in relation to the celestial sphere, being subject to a slow double-conical motion in a period of 26,000 years, caused by lunisolar attraction of the equatorial swelling." This movement, observed as early as 135 BC by the Greek astronomer Hipparchus of Nicaea (190–120 BC):

> translates into the spring equinox coming 21 minutes earlier [...] For this reason the phenomenon is referred to as the precession of the equinoxes. During the precession movement, the earth's axis undergoes oscillations leading it to follow at every point, other than the centre of the earth, a sinusoidal path over a period of 18 years; these secondary movements are known as nutations.
>
> (Strahler and Strahler, 1965, p. 56)

The instantaneous north pole is the point at which the imaginary line of the Earth's rotational axis passes through its surface. This point is not fixed but moves slowly towards the west (polar motion), due to a slight oscillation of the earth's axis during rotation; the "Chandler wobble", named after the astronomer Seth Carl Chandler who first observed it in 1891, is the outcome of the earth's geoid shape and internal movements underneath its crust, while an additional minor oscillation is determined by the seasonal redistribution of air and water masses. Predicted by the mathematician Euler in the eighteenth century, the Chandler wobble is an irregular circle some metres in diameter that is traced by the pole in a clockwise direction over about 14 months: "The general result of a preliminary discussion is to show a revolution of the earth's pole in a period of 427 days, from west to east, with a radius of thirty feet, measured at the earth's surface" (Chandler, 1891, p. 65).

At the centre of this irregular circle is the average rotation pole. Monitoring of the movements of the earth's axis was initiated in 1899 by the International Polar Motion Service; in 1987 the International Astronomical Union and the International Union of Geodesy and Geophysics set up the IERS (International Earth Rotation Service), which in 2003 became the Earth Rotation and Reference Systems Service:

> The variability of the earth-rotation vector relative to the body of the planet or in inertial space is caused by the gravitational torque exerted by the Moon, Sun and planets, displacements of matter in different parts of the planet and other excitation mechanisms. The observed oscillations can be interpreted in terms of mantle elasticity, earth flattening, structure and properties of the core-mantle boundary, rheology of the core, underground water, oceanic variability, and atmospheric variability on time scales of weather or climate.
>
> (Central Bureau, Federal Agency for Cartography and Geodesy, 2010[10])

The magnetic north pole (dip pole), as defined by William Gilbert (1600), is the point at which the terrestrial magnetic field is perfectly vertical, that is to say, where it forms an angle of 90° with the surface of the earth:

> The main problem with using the compass as an indicator of direction is related to the difference in position between the geographic North Pole and the magnetic North Pole. The magnetic North Pole is situated at around 70° N latitude, and 100°W longitude, near the Boothia Peninsula and Prince of Wales Island in the Northwest Territories of Canada. [...] the position of these poles varies continuously, which represents a further cause of uncertainty in the use of the compass.
>
> (Strahler and Strahler, 1965, pp. 42–43)

The position of the magnetic pole, which is measurable both directly and based on global models of the geomagnetic field, moves within a radius that varies from several to 100 km per day as a function of geomagnetic activity:

> It has been known since the seventeenth century that Earth is permeated by a magnetic field; the configuration of the lines in this force field may be visualized as being generated by a gigantic magnet at the centre of the earth; the north pole of this magnet may be imagined to be in the southern hemisphere and pointing towards the geographical South Pole, while conversely the south pole may be imagined to be in the northern hemisphere and pointing towards the geographical North Pole.
>
> (Meloni, 1993, p. 41)

Given that the terrestrial magnetic field is not perfectly dipolar, the two magnetic poles continuously change position and are not antipodal: currently, the magnetic north pole is located in Northern Canada, near Ellesmere Island (Korte and Mandea, 2008). The geomagnetic north pole, to complicate matters even further, is the point at which the approximately dipolar field of the earth's magnetosphere intersects with the surface of the planet. The difference between the magnetic and geomagnetic poles is based on the convention of reducing the earth's complex and varied magnetic fields to that of a dipolar bar magnet:

> the terrestrial magnetic field is caused by activity inside the planet and therefore external coefficients are not useful for representing it. The mathematical analysis of the potential of the earth's magnetic field [...] yields [...] precisely the potential of a magnetic dipole, which may be imagined as located deep within the Earth. The representative dipole does not perfectly coincide with the earth's rotational axis but forms an angle of about 12° with it. [...] The practice of confusing the geographical North Pole (intersection of the earth's rotational axis with its surface) and the geomagnetic North Pole (intersection of the representative dipolar axis with the northern hemisphere) is therefore only an approximation.
>
> (Meloni, 1993, p. 42)

A further complication from the physical point of view is that the magnetic pole located in the northern hemisphere corresponds to the south pole of a bar magnet, that is to say, the compass needle points to what is physically the earth's magnetic south pole, that is to say, towards the Arctic: "Note a potential cause of confusion regarding (geo-)magnetic poles: according to the physical definition, with respect to the direction of magnetic field lines, the poles on the northern hemisphere are south poles and vice versa. This is confusing in a geographic reference frame" (Korte and Mandea, 2008, p. 938). The geomagnetic pole, currently located near Thule, in Greenland, in contrast with the magnetic pole, is antipodal to, and moves simultaneously with, its southern counterpart. Recorded measurements from 1830 to date reflect a slowing down of dipolar movement, in contrast with the acceleration, from 1970 onwards, in the movement of the magnetic north pole, which is shifting up to 60 km/year in the direction of Siberia (ibid.); furthermore, analysis of materials from the ocean beds suggest a phenomenon involving cyclical reversals in polarity of the earth's magnetic field; the most recent polar shift is thought to have taken place some 780,000 years ago. The date, which is not currently possible to predict, of the next magnetic flip, as well as the present ongoing weakening of the terrestrial magnetic field that protects the earth from solar radiation, sometimes give rise to catastrophic speculation in the media regarding the future of life on our planet.[11]

North Pole is also the name, according to the Scientific American, of a town in Alaska (as well as a small community and Christmas theme park in the state of New York). Finally, we may add the northern pole of inaccessibility, that is to say, the geographical point most distant from any land mass, located at the centre of the Arctic Ocean's ice pack.

Lost on the map

> He had bought a large map representing the sea,
> Without the least vestige of land:
> And the crew were much pleased when they found it to be
> A map they could all understand.
> "What's the good of Mercator's North Poles and Equators,
> Tropics, Zones, and Meridian Lines?"
> So the Bellman would cry: and the crew would reply
> "They are merely conventional signs!
> Other maps are such shapes, with their islands and capes!
> But we've got our brave Captain to thank:"
> (So the crew would protest) "that he's bought us the best –
> A perfect and absolute blank!"
>
> (Lewis Carroll, *The Hunting of the Snark*, 1876)

What is a map? According to J.B. Harley: "The starting assumption is that maps constitute a specialized graphic language, an instrument of communication that has influenced behavioral characteristics and the social life of humanity" (Harley, 1987, p. 1). However, implicit in the map and its extraordinary importance as a tool of communication, is the paradox represented by the map itself: on the one hand, "a relatively simple iconic device" (ibid., p. 2)[12] and "physical artifact" (ibid., p. 1), on the other, the codification of "a complex language [...] whose properties we know very little about" (Harvey, 1969, p. 370). Harley argues that:

> Even after exhaustive scrutiny maps may retain many ambiguities, and it would be a mistake to think they constitute an easily readable language. Maps are never completely translatable. [...] maps are no less imprecise than written language. Although a key or legend may be provided, a line, a dot,[13] or a color, for instance, may have had several meanings, both manifest and latent, and it is unwise to assume that identical cartographic signs have similar meanings, or even a common origin, when found in different cultures.
>
> (ibid., p. 3)

Scale, the intrinsically approximative nature of cartographic communication, the choice and use of symbols to highlight selected elements, the dynamic and shifting nature of the territory, culture and the cartographer's own vision of the world, all contribute to making the language of maps far removed from any pretension to "photographing" reality, or to providing "a mirror of nature", to use the language of Rorty (1979). Furthermore, the map influences our way of thinking about the world, given that none of us can see the world in its entirety and therefore the perspective from which it is represented creates perceptions of closeness and distance that are not necessarily real.

In Harley's view, the map is above all an instrument of power: "Power comes from the map and it traverses the way maps are made. [...] The map is a silent arbiter of power" (Harley, 1989, p. 12). As implied by Korzybski's well-known remark – later revisited and critiqued by Bateson – that "the map is not the territory, and the name is not the thing named", producing a map involves processes of transforming, decoding, and interpreting reality:

> What is on the paper map is a representation of what was in the retinal representation of the man who made the map; and as you push the question back, what you find is an infinite regress, an infinite series of maps. The territory never gets in at all. [...] Always, the process of representation will filter it out so that the mental world is only maps of maps, ad infinitum.
>
> (Bateson, 1972, p. 460)

Thus for Harley, following Foucault and Derrida, cartography is discourse, or an expression of political power and as such it transmits a given vision of the world, making cartography the "business of the state": "Cartography deploys its vocabulary accordingly so that it embodies a systematic social inequality. The distinctions of class and power are engineered, reified and legitimated in the map by means of cartographic signs. […] Much of the power of the map, as a representation of social geography, is that it operates behind a mask of a seemingly neutral science" (Harley, 1989, pp. 6–7). In *Abysmal. A Critique of Cartographic Reason*, "a minimalist guide to the landscape of western culture", Gunnar Olsson narrates his own journey, equipped with personal map and compass, through: "A chaosmos in which categories ride the surf, traces are erased, drops turn into clouds, clouds change into drops" (Olsson, 2007, p. 39). En route, Olsson engaged with the art of cartography: "What is learned along the way is not only that every map is simultaneously a picture and a story but that every map is a record of mistranslation, hence a major confrontation with the limits of representation". Ultimately, Olsson concludes that every cartographer is:

> ([.] equally skilled in the art of rhetoric, some claim in the art of lying), […and] knows that in order to make a map all that is needed are three concepts. No more and no less than
>
> . a point which remains at rest;
> . a line better known as a scale or a coordinate-net; and
> . a plane onto which the emerging picture/story is projected.
>
> And so is it that despite all its complexity mapping is essentially an exercise in the geometry of triangulation, triangulation itself an exercise in the politics of power, the latter a dance with the semioticians' three-cornered figure of thought, symbol and referent.
>
> (2007, p. 126)

In this dance posited by Olsson, maps always require interpretation. Queequeg, the heroic "cannibal" in Moby-Dick, "was a native of Kokovo, an island far away to the East and South. It is not down in any map; true places never are" (Melville, 1981, p. 60). But if true places never appear on maps, are we to suppose that maps always lie? Monmonier's answer to this is that

> not only is it easy to lie with maps, it's essential. To portray meaningful relationships for a complex, three-dimensional world on a flat sheet of paper or a video screen, a map must distort reality. As a scale model, the map must use symbols that almost always are proportionally much bigger or thicker than the features they represent. To avoid hiding critical information in a fog of detail, the map must offer a selective, incomplete view of reality. There's no escape from the cartographic paradox: to present a useful and truthful picture, an accurate map must tell white lies. Because most map users willingly tolerate white lies on maps, it's not difficult for maps also to tell more serious lies. […] Yet cartographers are not licensed, and many mapmakers competent in commercial art or the use of computer workstations have never studied cartography. Map users seldom, if ever, question these authorities, and they often fail to appreciate the map's power as a tool of deliberate falsification or subtle propaganda.
>
> (1991, p. 1)

The implicit or intentional lies contained in maps, as well as mistakes in interpreting them, can cause disorientation, at times with serious consequences. On 24 November 1940, a lack of experience in map-reading led a British bombardier to lose his way and mistakenly strike Marseilles instead of Genoa, 200 miles distant. A report on the mishap in the American Nazi propagandist journal *Facts Ill. Review* (30 November 1940) showed the distance on the map between the two cities, instrumentally citing the incident as proof of foul play on the part of the English: "British Bombings – A Record of British Truthfulness" (ibid., pp. 103–104). If maps always lie, then Monmonnier suggests that "The wise map user is thus a sceptic, ever wary of confusing or misleading distortions conceived by ignorant or diabolical map authors" (ibid., p. 157).

The map expresses the worldview of the central power that has made it: "Mapping soon became the business of the state: cartography is early nationalized. The state guards its knowledge carefully: maps have been universally censored, kept secret and falsified. In all these cases maps are linked to what Foucault called the exercise of 'juridical power'" (Harley, 1989, p. 11). For "reasons of security" and in the name of protecting the state, maps may contain deliberate errors, designed to disorient the enemy and avoid giving away information about sensitive targets. The systematic falsification of maps can become a consolidated defence tactic. For 50 years, the Soviet Union engaged in the modification of its maps for strategic reasons. From the 1930s onwards, rivers, buildings, streets, borders, coastlines, towns, roads and railway lines in the USSR were either omitted from maps or their positions changed as instructed by the secret police force, the N.K.V.D. – which had responsibility for map-making – with a view to discouraging air raids and withholding information from foreign intelligence services. During the Second World War, the inaccuracy of maps in the USSR contributed to Germany's defeat in Russia, leading the German forces to either get lost or wind up stuck on muddy roads that had been misleadingly represented on the map. Alexander Stahlberg, an official with the 12th Panzer Division during Operation Barbarossa, complained about the Soviet maps, describing them as "fit only for use as lavatory paper". General Blumentritt commented that they "in no way corresponded with reality". On these maps, the main roads were marked in red and appeared to be numerous, encouraging a deceptive confidence in what was to come. The reality, however, was often nothing more than "sandy tracks" (Stahel, 2009, p. 235). The use of cartographic error to disorient the enemy may carry over to the manipulation of signage on the ground. During the "Prague Spring", Czechs and Slovakians deployed every possible strategy to halt the Soviet occupation via acts of civil disobedience. In Prague, Bratislava, and other cities, hundreds of street signs were tampered with, taken away, destroyed or moved in order to confuse the invaders and send them in the wrong direction (Vogt, 1971, p. 67).[14]

Soviet cartographic misinformation continued throughout the Cold War, with maps and atlases deliberately offering false information and intentionally leaving out key sites and infrastructure or marking them in the wrong position, in keeping with the Soviet Map Distortion Policy launched in 1958. The city Logashkino, on the coast of the East Siberian Sea, either appeared in different locations or disappeared altogether on maps dated between 1939 and 1969.[15] Again Monmonnier quotes an unsigned journal article in the *The Military Engineer* as stating: "Apparently there is such a town, but whether it is on the seacoast or on a river, or neither, is a matter of uncertainty when based on the work of the Soviet cartographer" (1991, p. 117). Paradoxically, the use of such disorientation tactics intensified from the late 1960s onwards, in parallel with the introduction of satellites across the United States. Even tourist maps began to feature errors and omissions, often failing to provide a scale or identify key buildings such as the highly visible headquarters of the KGB in Dzerzhinski Square in Moscow, today Lubyanka square.

Clearly, using maps to misinform is not an exclusively Soviet prerogative. Sensitive sites have also been camouflaged or hidden in the British *Ordnance Survey* system and in the cartography of other countries such as Greece or Italy. Emanuela Casti attributes a key role to cartographic misinformation and the mistaken interpretation of a sketched map of the Tigray region in Italy's 1896 defeat at Adwa in Ethiopia, during the first Italo-Ethiopian war contested between the Ethiopian empire and the Kingdom of Italy:

> All the misunderstandings and errors [...] seem to have been due to the fact that it was impossible to match the territory with the information provided in the sketch; or rather, the incongruence between that which was reproduced by the map and that actually presented by the territory was received by attributing a credibility to the map unjustified by the directly available evidence.
>
> (Casti, 1998, pp. 170–171)

Enrico Squarcina's perspective on this phenomenon is that:

> It is the map that constructs the world and not vice versa. Even when our past experience teaches us to be on the alert and makes us doubt that it is the other way around, it is difficult to turn our backs on the power conferred by the map, it is difficult to doubt the power issuing from the map. Only a critical spirit such as Emilio Lussu is capable of the diffidence of one who has learnt from experience to question both power and its map: "Do not trust maps. Otherwise, you will never find your regiment again. Believe me, as an old army official with a long career behind him. I fought the entire campaign of Africa [First Italo-Ethiopian War]. We lost at Adwa, because we had a few maps. And so we finished up in the west instead of going east. Something along the lines of attacking Venice instead of Verona. In mountainous regions, maps are only intelligible to those who are familiar with the region, because they were born or have lived there. But those who already know the territory do not need maps."[16]
>
> (Squarcina, 2009, p. 89)

In the United States, and in many other countries, large-scale maps tend to omit details that are politically "awkward", as in the well-known case of Love Canal:[17] "American maps often omit information that might embarrass industrial polluters or local officials" (Monmonier, 1991, p. 121). Similarly, monuments, churches, and museums are marked on maps, but not poor districts, areas with a high rate of crime, or features and infrastructural elements of dubious aesthetic appeal:[18] "By omitting politically threatening or aesthetically unattractive aspects of geographic reality, and by focusing on the interests of civil engineers, geologists, public administrators, and land developers, our topographic 'base maps' are hardly basic to the concerns of public health and safety officials, social workers, and citizens rightfully concerned about the well-being of themselves and others. In this sense, cartographic silences are indeed a form of geographic disinformation" (ibid., p. 122).

The aim of a map is to transmit a certain message to a certain public. Lewis Carroll's absolutely "blank" sea chart in *The Hunting of the Snark* is easy for the crew to read because it is devoid of conventional signs and therefore anybody can use it. In the words of Harley: "Again we ought to dismantle the arbitrary dualism between 'propaganda' and 'true', and between modes of 'artistic' and 'scientific' representation as they are found in maps. All maps strive to frame their message in the context of an audience. All maps state an argument about the world and they are propositional in nature. All maps employ the common devices of rhetoric such as invocations of authority (especially in 'scientific' maps) and appeals to a potential readership through the use of colors, decoration, typography, dedications, or written justifications of their method. Rhetoric may be concealed but it is always present, for there is no description without performance" (Harley, 1989, p. 10). Maps, especially when thematic, tend to make the simple complex, making it difficult to identify in the real world what is symbolized on the map. The art of cartographic representation helps to convey the message that one wishes to transmit, through symbolic language and the choice of contour lines, shading, and colours. This art, and the cartographer's degree of skill in applying it, can also confuse the map-reader. Colour and contrast may make a map attractive but transmit wrong information to those who are not expert in the territory represented:

> For several centuries, cartographers have exploited and encouraged such associations as green with vegetation, blue with water, red with high temperatures, and yellow with a desert environment, and where the context is correct and appropriate, these associations promote efficient decoding. [...] Confusion is likely if the reader associates white with snow, green with abundant vegetation, and yellow or brown with desert – much of the world's tundra is close to sea level, many lowland areas are deserts, and many upland areas have substantial forests or grasslands.
>
> (Monmonier, 1991, p. 154)

Maps show us unknown and mysterious places. Charlie Marlow, at the opening of Conrad's *Heart of Darkness*, tells of his fascination with maps: "Now when I was a little chap I had a passion for maps. I would look for hours at South America, or Africa, or Australia, and lose myself in all the glories of exploration. At that time there were many blank spaces on the earth, and when I saw one that looked particularly inviting on a map (but they all look that) I would put my finger on it and say, 'When I grow up I will go there.' [...] But there was one yet – the biggest, the most blank, so to speak – that I had a hankering after" (Conrad, 2006, p. 11). Are there still "blank spaces" on the face of the Earth or do present-day maps display the whole of reality? Despite remarkable advances in information technology and contemporary cartography, the pace of change in the earth's surface due to both human (urban development) and non-human (erosion, deforestation) processes means that even the most recent maps require constant updating if they are not to become obsolete. As commented by Manik Gupta, former group product manager at Google Maps: "The very moment you build a perfect map of the world is the moment it goes out of date [...] The real world will always be a little bit ahead of how we represent it, because change is constant" (Gupta, in Nuwer, BBC, 2014).[19]

Cartographic errors have persisted into recent times, presenting us with inexistent places and routes to reach them. One of the most infamous of such mistakes concerned the so-called Mountains of Kong, a mountain range supposedly located along the western seaboard of Africa, which were erroneously mapped and described by the British cartographer James Rennel in 1798 but continued to feature in atlases and maps for over a century. In 1890, the Mountains of Kong still appeared on a map of Africa by Rand McNally. For over two hundred years, up to the mid-1800s, California was represented on maps as an island in the Pacific, despite numerous failed attempts to circumnavigate it. In 1875, Sir Frederick Evans succeeded in "cleaning up" the Pacific by having 123 inexistent Pacific islands removed from the British Admiralty Charts (only three of these actually did exist). The fictitious islands included Morris Island (near Hawaii) and New South Greenland (towards the Antarctic), both the imaginary creations of navigator Benjamin Morrell who described them in detail in the 1820s and 1830s. New South Greenland continued to feature in atlases up to 1915 (Garfield, 2012). Still more mysterious is the story of the "phantom" isle known as Sandy Island in the Coral Sea, near New Caledonia, "discovered" by James Cook in 1772, and later shown to be inexistent but left on maps. Google Maps only removed it from their Internet mapping service in 2012 (Seton et al., 2013).

How can we avoid getting lost in the ambiguity of the map? Maps are above all a language, the translation onto a sheet of paper of the extraordinary complexity and dynamism of a territory, a translation that implies distortion, error, and fantasy. As we have seen, a map should always be read as a particular view of the world and as an instrument of power that is influenced by the needs of the authority that produced it; maps may even become instruments of political propaganda and the information contained in them may be manipulated to suit strategic objectives, whether we are talking about a military requirement to disorient enemies and conceal sensitive targets, or a civilian requirement to highlight or hide places that are more or less attractive. Though intended to be a precise aid to orientation, cartographic representation is inevitably approximative and may even contain imaginary places or mistakes. From a technical point of view, symbols, colours, contour lines, and shading must be expertly chosen if the final product is not to confuse and disorient the inexperienced map-reader. As well as acting as an information tool, maps can inspire passion, a passion that springs from the desire to learn the truth about faraway and unknown places or simply from the desire to imagine and represent new places born of one's fantasy. Whether the information contained on a map is true or false, it remains an extraordinary medium of communication. Thus, we may appropriately give the last word to Wittgenstein:

> It is easy to imagine and work out in full detail events which, if they actually came about, would throw us out in all our judgments. If I were sometime to see quite new surroundings from my window instead of the long familiar ones, if things, humans and animals were to behave as they never did before, then I should say something like "I have gone mad"; but that would merely be an expression of giving up the attempt to know my way about. […]. But the important thing about this for me is that there isn't any sharp line between such a condition and the normal one.
>
> (Wittgenstein, 1967, p. 393)

Notes

1 See Brandt (1980).

2 Thus, the title of the French film directed by Dany Boon *Bienvenue chez les Ch'tis* (2008) as well as its Italian translation *Benvenuti al Sud* [Welcome to the South] (Medusa Film 2010), is a play on national stereotypes.

3 The anthropologist Caroline Humphrey in the course of in-depth fieldwork, observed the orientation system of the inhabitants of Inner Mongolia: "There are several 'mental maps' by which Mongols orient themselves. […] The Black Dog is a notional cosmological creature lying on its side, which moves round the directions (north, east, south, west) as it were like a timepiece, so that its mouth gradually moves from one direction to another during the seasons of the year. The direction of the Dog's Mouth is always a bad direction to take. The dog's chest faces a good direction, we were told, but the tail is unpredictable […] People in rural Urad, in fact throughout rural Inner Mongolia, always know the direction of the Black Dog's Mouth in any given period of the year. The Black Dog system applies to everyone, but the other system, that of the Eight Seats (M. *naiman suudal*) applies to people differently according to their date and time of birth, their age and their sex. The Eight Seats are astrological sites linked to the eight cardinal directions; they determine which will be auspicious and inauspicious directions for individual people to take at a given time. The two systems of orientation just mentioned are notably symmetrical, yet the physical land people live in cannot be so" (Humphrey, 2001, p. 64).

4 "The dead have their heads placed towards the North, the living face the South" (Confucius).

5 See also Juhel (2013).

6 … mov'd a voice, That made me seem like needle to the star, In turning to its whereabout.

7 http://www.treccani.it/vocabolario/polo1. Accessed 2 April 2017.

8 http://www.ngdc.noaa.gov/geomag/GeomagneticPoles.shtml. Accessed 2 April 2017.

9 See also Davidson (2016).

10 http://www.ggos-portal.org. Accessed 2 April 2017.

11 See, for example, the disaster film *The Core* (2003) directed by Jon Amiel.

12 See also Robinson and Petchenik, *Nature of Maps,* Chapter 3 (footnote 4), quoted by Harley (1987).

13 The exponent of Land Art, Robert Smithson, observed that: "The dot evades our capacity to find its center. Where is the central point, axis, pole, dominant interest, fixed position, absolute structure, or decided goal? The mind is always being hurled towards the outer edge into intractable trajectories that lead to vertigo" (Smithson, 1996, p. 94).

14 From: "Geschichte: Prager Frühling – eine Verschwörung?" Kilian Kirchgeßner (15.12.2008). Source: http://www.dw.com/de/der-prager-frühling-eine-verschwörung/a-3876683. Accessed 2 April 2017.

15 Source: *The Military Engineer*, No. 410, November–December 1970.

16 See Lussu (1945, p. 27).

17 "The 1980 map not only shows no trace of the filled-in canal but ignores the area's tragic history: dumping continued until 1953; developers filled in the canal and built homes there in the 1950s; the city built a public elementary school across the filled-in canal in 1954; chemical seepage spread up to the surface and laterally into the basements of nearby homes. After people and pets became ill in the early 1970s, soil analyses revealed abnormally high concentrations of chlorobenzene, dichlorobenzene, and toluene; and in 1978 the New York health commissioner declared a state of emergency and relocated 239 families. Although both federal and state mapping agencies might contend that topographic maps should show only standardized sets of readily visible, more-or-less permanent features, such assertions seem hypocritical when these agencies' maps routinely include boundary lines, drive-in movie theaters, and other elements far less important to human health" (Monmonier, 1991, pp. 121–122).

18 The Missing Maps Project – organized by the Red Cross, Doctors Without Borders, and the Humanitarian Open Street Map Team – recruits volunteers to fill in cartographical blanks in the developing world.

19 Rachel Nuwer, *The Last Unmapped Places on the Earth*, 28 November 2014, http://www.bbc.com/future/story/20141127-the-last-unmapped-places.

2 (Dis)orienting oneself in thinking

The art of losing isn't hard to master[1]

On the Canadian website *gettinglost.ca – understanding human navigation* set up by NeuroLab, the cognitive neuroscience laboratory of the University of Calgary directed by Giuseppe Iaria, we read that:

> Finding our way in spatial surroundings is one of the most complex behaviours that humans perform on a daily basis. It allows us to reach a target location while navigating through a range of environments, from feature-rich cities to natural landscapes that stretch the horizon. This complex phenomenon relies on different orientation strategies that individuals may use either selectively or in combination with one another.
>
> (How We Orient, Fordulation, 2011, www.gettinglost.ca, accessed 2 April 2017)

But what does "getting lost" mean from a cognitive viewpoint? In the work *Why People Get Lost. The psychology and neuroscience of spatial cognition*, an examination of "the relationship between the brain and the ability to find one's way around an environment", psychologist Paul Dudchenko explains that "being lost means being unable to find one's way. In neurological studies, this inability to navigate large-scale spaces is sometimes referred to as *topographical disorientation*" (Dudchenko, 2010, p. 4). According to John J. Rieser, a psychologist with a special interest in how wayfinding works in the visually impaired: "The phrase *losing one's way* typically means that one does not know the directions and distances needed to get back to a starting point or other useful destination while one is under way" (Rieser, 1999, p. 172). This definition introduces the concepts of spatial orientation, and of changes in geometric relations, distances and direction. In the words of Reginald G. Golledge: "Anyone experiencing even momentary disorientation and lack of recognition of immediate surroundings has experienced the uncertainty of being 'lost'. This state occurs when the wayfinding process being used to guide travel fails in some way" (Golledge, 1999, p. 5). Thus, if we are to interpret the experience of disorientation in human beings, it is essential to have developed an understanding of the wayfinding process, in terms of how our cognitive structures interact with environmental cues. Wayfinding, according to Golledge, consists in "the cognitive and behavioral abilities of human and nonhuman species to find a way from origin to a destination. The way may involve following existing paths or tracks or free-ranging movement" (Golledge, 1999, p. 24). During the trajectory, the distal destination is not directly visible to the traveller. Golledge concludes that: "An element of uncertainty is a factor in every effort. Ultimately, successful wayfinding is reflected in the traveler's ability to achieve a specific destination within the confines of pertinent spatial or temporal constraints and despite the uncertainty that exists" (ibid., p. 47). The spatial abilities of both humans and nonhumans have been studied within a range of disciplines[2] for over a hundred years now. The existing substantial body of experimental research and field observations of people and animals prompts Dudchenko to conclude that "humans aren't terribly accurate in returning to a start point in the absence of visual landmarks. They exhibit errors of 20–30° after travelling relatively short distances, and even show a 20° error when turning around after walking in one direction" (Dudchenko, 2010, p. 82).

How do human beings orient themselves? Are we equipped with some kind of "sixth sense", highly developed or otherwise, for finding our way? The scientific literature reviewed by Dudchenko suggests that we are not, particularly given our performance in environments lacking readily identifiable landmarks: "These findings provide little support for the view that humans are able to keep track of their directional orientation by following larger-scale passive movement in the absence of visual information" (ibid., p. 85). Neither do humans appear to be endowed with a magnetic sense of direction: "Over longer distances, there isn't consistent evidence for a magnetic sense. So, one of the reasons we get lost is because, in the absence of vision, our capacity for keeping track of our own orientation or that of a target location is limited" (ibid., p. 92). Field data collected by a number of scholars and the large body of literature on the topic (see Dudchenko, 2010; Golledge, 1999) suggest that human beings, differently to other animal species such as rodents, rely principally on visual cues to orient themselves – hence their tendency to seek out visible landmarks – with the other senses making a minimal contribution to orientation. Allen remarks that: "Although various species are sensitive to distal visual (including solar, lunar, and stellar), tactile (including wind and water currents), geomagnetic, and olfactory information, humans tend to rely most heavily on visual or, in case of visual impairment, on auditory, vestibular, and proprioceptive information" (Allen, 1999, p. 49).

Systematic analysis of the behaviour of lost persons dates back to the 1970s. It has been spearheaded by search and rescue experts from different countries, and has resulted in the publication of a significant body of studies and statistical data. Pioneering work in this field was carried out by William Syrotuck, who in 1976 published a study of 229 persons reported lost in the states of New York and Washington. Robert Koester, in *Lost Person Behavior*

(2008), identified 41 categories of missing persons, based on the *International Search & Rescue Incident Database* (ISRID), which contains data on 50,000 SAR incidents.[3] Among these categories, children, the elderly, and persons with physical or mental impairment display particular difficulty in finding their way: "Disorientation results not from the incapacity to perceive the information but the incapacity to accord it a meaning, as can be observed with the young child who has not completed the necessary learning or in the old people suffering from neurodegenerative diseases" (Cadet, 2010, p. 206). Psychologist Kenneth Hill, a member of the *Halifax Regional Search and Rescue Team*, has identified a set of standard behaviours adopted by people who get lost in wooded areas. These include route traveling (taking alternative paths), route sampling (trying out different paths from a known intersection), seeking a better view of the surrounding area by going to higher ground, and backtracking. Some of the less effective methods noted by Hill are random traveling (following the easiest trajectory) and direction traveling, in which the lost person chooses one particular direction and pursues that, ignoring paths or other landmarks encountered along the way. An even riskier option is to rely on popular wisdom, such as "all streams lead to civilization", which has no scientific basis and in Nova Scotia would probably lead one into an insect-infested swamp. The most effective strategy, according to Hill, is to stay put and wait to be rescued, although typically very few lost persons choose this method. Stopping and waiting can also help to mitigate anxiety and panic, which many scholars note to be a typical component of the disorientation experience. Golledge comments: "There is evidence that humans and other species make errors and develop strategies to compensate them [...]. When using an environment, certain types of choices (e.g., with respect to travel) must be made. Humans making an incorrect choice can become lost and suffer the psychological and physiological discomfort associate with that state of being" (Golledge, 1999, pp. 42–43). In Hill's research on a sample of 800 persons who got lost in the woods of Nova Scotia, only two opted to stay put. Hill concludes: "However, it is clear from this research that having a good 'sense of direction' is based on the ability to take advantage of environmental cues, including feedback from one's own body movements, rather than a mysterious sixth sense" (Hill, 1998, p. 4).

You are here

Let us now consider an example of how people ordinarily experience (dis) orientation when they need to navigate an unfamiliar space in the course of their everyday lives. On exiting the subway of a strange city, a tourist stops to look at a panel displaying a map of the area. Although on setting out he had a vague grasp of where his destination, say for instance a museum, was located, he has completely lost his bearings during his underground journey. Large cities "full of names and memories" (Lynch, 1960, p. 124) offer many landmarks and points of reference, but for this same reason, may be experienced as chaotic and overstimulating and therefore difficult to interpret. Kevin Lynch goes so far as to compare our behaviour in an unknown city to that of a person with brain damage, explaining that "[t]he terror of being lost comes from the necessity that a mobile organism be oriented in its surroundings" (ibid., p. 125). Underground rail travel is particularly likely to compromise the workings of what Erik Jonsson, the Swedish author of *Inner Navigation* (2002), refers to as our "inner compass". The diagram of the underground network is not of much assistance to the tourist either, in that it is only based on the sequence of stops and the direction of the line. Although our tourist may know the name of the station at which he boarded the subway and that of the one where he is now, once above ground he cannot tell what direction he has come from.

In New York City, precisely to help travellers overcome this sense of confusion and bewilderment, pavements outside selected underground stations (e.g., *Grand Central Terminal*) have been experimentally equipped with compass decals[4] indicating north, south, east, and west, as well as the names of the nearby streets. The New York Metro, as those who travel on it regularly are well aware, is a truly labyrinthine network that developed from the 1940 unification of three earlier urban rail networks,[5] and therefore it is common for passengers – even locals – to find themselves completely disoriented on exiting a station and stepping onto a big street without easily identifiable points of reference. Travelling by underground can be disorienting in any case (see § 8.7). Eric Jonsson, despite the strong orientation skills that he had developed in the forests of his native Sweden, describes how totally confused he felt on exiting the Paris Metro, on arrival in France after a long journey: "My sense of direction must have been confused during the underground ride, so that when I emerged into the street at the Chateau d'Eau station, my 'automatic pilot' jumped to the wrong conclusion and made me feel that north was south. [...] A most humiliating feeling for a Swedish forest hiker, who prided himself on his good sense of direction" (Jonsson, 2002, p. 17).

Unless the tourist in our example has a smartphone on which to consult Google Maps or a GPS device that can suggest a route for him to follow, at this point he only knows that the museum he is looking for is nearby. From a cognitive psychology perspective: "Human navigation involves several stages of information processing, for example, acquisition, coding, and integration of sensory information, as well as processing spatial information in order to localize oneself in relation to the environment and the objects it contains. Cognitive psychology makes extensive use of reaction times [...] and errors as performance measurements in order to infer the elements of mental organization" (Amorim, in Golledge, 1999, p. 153). The tourist's current position corresponds to a circle or dot on the map that usually bears the label "*You are here*", a choice of wording that is anything but random. As Amorim explains: "The use of 'you-are-here' maps implicitly suggests that *viewer-centered* (VC) frames of references are easier to use than *object-centered* (OC) ones [...] In Paris, a map indicating 'the Eiffel tower is here' would probably make it more difficult for the tourist to find his or her current position with respect to the monument" (ibid., p. 157). To find a museum on the map, the tourist is obliged to carry out a series of complex mental operations: first, he must interpret the symbolic deixis inherent in the personal pronoun (*you*) by identifying himself as the intended referent, despite the fact that it is used generically and does not refer to exclusively to him. He must then relate the adverb of place *here* to the iconic representation of his position by means of the circle or dot; finally, he must put these two data items (personal pronoun + adverb of place) together to extract the new piece of data "I am here = at this precise point", where I, in this case, does not correspond to a psychological or social dimension but rather to the position of his body, situated in a place as yet to be precisely defined, in that it does not belong to the two-dimensional street map which he is currently viewing but to three-dimensional real space: "Such a hypotetico-deductive strategy may allow us to match our three-dimensional views with the two-dimensional representation provided by the map" (Dudchenko, 2010, p. 6).

The cognitive map

The tourist must now draw on his internal sense of direction to organize the spatial information he has acquired into a "mental map" and decide on the route to take to the museum. The term "mental map" or "cognitive map", understood as an "internal spatial representation of environmental information" (Golledge, 1999, p. XIV) was used for the first time by Edward Chase Tolman (1948) in his ground-breaking study on the use of spatial information in animals. The first behaviourist experiments with rats were conducted in the early 1900s in mazes modelled on the famous Hampton Court Maze in Surrey, England. In his own research, Tolman found that rats were able to construct a mental map based on their exploration of the internal environment of a maze. Tolman rejected the reductionist stimulus-response paradigm, arguing that: "In the course of learning something like a field map of the environment gets established in the rat's brain. [...] And it is this tentative map, indicating routes and paths and environmental relationships, which finally determines what responses, if any, the animal will finally release" (Tolman, 1948, p. 193). Constructing a mental map is an individual process, and the time required to do so varies as a function of our capacity to store, access, and retain the spatial information pertaining to a given dynamic environment. Our spatial skills and repertoire of wayfinding strategies depend – among other factors – on our prior experience, the characteristics of the environment we are navigating, and our age; such factors influence how we interpret position, distances, or points of reference.

From a geographical perspective, Tolman's "rat in the labyrinth" experiments were taken up by Jan Pailhous in *La représentation de l'espace urbain. L'exemple du chauffeur de taxi.* In Pailhous' opinion: "The similarities between the rat in its labyrinth and the taxi-driver in his city go much further than a simple play on words. Indeed, although the labyrinths utilized are for the most part simple, or even the most basic possible (T-shaped), it is no less the case that a goal that cannot be perceived must be reached by means of an operation of movement and that this very quickly raises the issue of the role played by cognitive processes" (Pailhous, 1970, p. 25). The taxi driver is a particularly interesting subject when it comes to investigating the orientation strategies that people adopt to navigate a large city and how they acquire knowledge of the complex space that is the urban network. In more recent studies, Maguire and colleagues (2000) and Spiers and Maguire (2006) have analyzed the brains of London taxi drivers, who are required to be familiar with the positions of the city's 25,000 streets and thousands of sites of interest as a prerequisite for obtaining their licence. In human beings, greater activity may be observed in the hippocampus of individuals engaged in the acquisition of spatial knowledge, and research suggests that this ability may be improved through practice. According to Maguire and co-researchers, the posterior hippocampus is significantly more developed and larger in London taxi drivers than in control subjects: "The volume of the hippocampus in nonhumans is known to vary as a function of the demands placed on spatial memory [...]. Similar effects have been reported in licensed taxi drivers in London (UK)" (Maguire et al., 2006, p. 1091). In relation to Parisian taxi drivers, in his essay on industrial psychology Pailhous further comments that

> One of the key characteristics of urban space is the requirement for all operators to move towards goals that they cannot perceive. From this perspective, the taxi-driver is a privileged operator; he is required to memorize, not just basic and frequently taken routes, but the entire urban network; and, what is more, he must know all the street names [...] Travelling around a big city requires operators to use a map, which we might view as an "aid to work" given that it prepares or facilitates the journey to be made.
>
> (1970, p. 8)

Once again, the implication here is that we construct our own personal image of the environment – the taxi driver's *plan* – and refer to it to organize our movement towards a goal and to devise our wayfinding strategies. According to Lynch: "The creation of the environmental image is a two-way process between observer and observed" (Lynch, 1960, p. 131). In order to move about the absolute space surrounding us, we must necessarily have formed a mental representation of it: "This representation [...] is prone to an accumulating error when we travel in an ambiguous environment or are inattentive. If uncorrected, we become lost" (Dudchenko, 2010, p. 7). Kevin Lynch adds:

> The environmental image has its original function in permitting purposeful mobility. [...] Way-finding is the original function of the

> environmental image, and the basis on which its emotional associations may have been founded. But the image is valuable not only in this immediate sense in which it acts as a map for the direction of movement. [...] In this way it is like a body of belief, or a set of social customs: it is an organizer of facts and possibilities.
>
> (ibid., pp. 125–126)

Our mental images may be organized in a variety of ways. Features of our surroundings, myths, symbols, and abstract reference systems can all help us to order our "consciousness of a familiar environment" and select points of reference, or landmarks, with which to construct a cognitive map. For Golledge: "Landmarks are the bricks of our spatial knowledge" (1999, p. 154). In finding our way along a route, we may follow a range of strategies, all of which are based on identifying these landmarks, that is to say: "the points of reference considered to be external to the observer, [...] simple physical elements which may vary widely in scale. [...] the key physical characteristic of this class is singularity, some aspect that is unique or memorable in the context. Landmarks become more easily identifiable, more likely to be chosen as significant, if they have a clear form" (Lynch, 1960, p. 78). What psychologists refer to as spatial cognition varies as a function of whether we choose reference points that are *egocentric*, that is to say, related to our own body (right, left, in front of, behind) or *allocentric*, that is to say, defined for absolute space, relative to other objects, such as the cardinal points (Dudchenko, 2010). O'Keefe and Nadel explain that "Orientations can be interpreted as sets of instructions for aligning the body axis relative to some other axis. In the simplest case this involves a rotation relative to the present orientation, i.e. a simple turn and continuation straight on" (O'Keefe and Nadel, 1978, p. 82). Egocentric reference systems define an observer's position in relation to his or her own body and offer a subjective perspective that depends on the observer's current orientation. Allocentric or absolute systems of reference are based on the environment, are not associated with a given perspective and are independent of the observer's body (Kosslyn, 1994; Paillard, 1991; McNamara et al., 2003).

Thus we have two different ways of knowing where we are: "One [...] is a cognitive map – an allocentric memory for the spatial layout of a given environment. The second is an egocentric updating of one's current position relative to landmarks in the current environment" (Dudchenko, 2010, p. 91). Memorizing a sequence of elements, points of reference and the shape of the surrounding environment can help us to correct our position when we lose our bearings: "The important feature of this process is that the cognitive system is not just photographing what is out there in the environment, but it is actively constructing the external and internal networks that make the cognitive map" (Stern and Portugali, 1999, p. 108). Human beings tend to organize environmental information using an internal orientation mechanism that allows them to navigate space: "This mechanism may work seamlessly in most situations, but may become detached from reality when we interrupt our contact with the environment by sleeping, or by attending to one aspect of the environment (for example a newspaper we might be reading) while being passively transported on a train or a bus" (Dudchenko, 2010, p. 5). This explains why the tourist in our earlier example, if he is not concentrating, finds travelling on the underground to be a particularly disorienting experience. In line with the two proposed modes of finding one's way, Dudchenko argues for distinguishing between two different ways of losing our bearings: misorientation and disorientation. In the case of misorientation:

> One may retain knowledge of the spatial relations between landmarks that are out of view, but one's heading relative to this representation is misaligned. [...]. Disorientation, in contrast, would refer to a breakdown in the map itself. The knowledge of the spatial relationships themselves becomes unstable. [...] So, one of the reasons we get lost is because, in the absence of vision, our capacity for keeping track of our own orientation or that of a target location is limited.
>
> (Dudchenko, 2010, p. 92)

Finding oneself "turned around" and walking straight in circles

One particular experience of misorientation and the deep distress that it can cause is the feeling of being "turned around" described by many authors and first mentioned in the context of a discussion on instinct among Charles Darwin and others in the journal *Nature* (see Chapter 5). In a 1873 contribution to this exchange, Henry Forde wrote about hunters in West Virginia who had experienced this type of *retournement*: "A kind of seizure; that they may 'lose their heads' all at once, and become convinced that they are going in quite the contrary direction to what they had intended, and that no reasoning nor pointing out of landmarks by their companions, nor observations of the position of the sun, can overcome this feeling" (Forde, 1873, p. 463). The topic was revisited by Binet in 1894. In his *Reverse Illusions of Orientation*, he reported accounts of hallucinatory experiences that featured a "reversal" of orientation in ordinary travellers or hunters: "The illusion is accompanied almost always by distress and anxiety" (Binet, 1894, p. 349). Binet cautioned that such phenomena had "not yet been studied methodically", although they had been observed by many scholars and labelled as "illusions or hallucinations of orientation" which could be experienced by day or by night. One of Binet's informants (Observation 1 by H. Beaunis) stated that: "I cannot better express the sensation which I felt than by the expression 'being turned around' (*renversement*). It seemed to me that the landscape turned itself about, and I travelled in a direction opposite to that which I followed on leaving Paris" (ibid., p. 339). Another subject (Observation II by M.J. Passy) elaborated still further: "My confusion was extreme, and I found myself for some time in a state of complete disorientation, and unable to comprehend the situation. […] What we have then in this particular case is a completely false orientation. All the objects and streets occupy positions exactly opposite to those which they should" (ibid., p. 340). The phenomenon of disturbances to orientation was also recognized by Viguier, who in *Le sens de l'orientation et ses organs chez les animaux et chez l'homme* (1882) described "a sort of vertigo of direction" giving rise to a general sense of malaise and disquiet, along the lines of *renversement*. This sensation could either suddenly disappear or fade away gradually. Lund later revisited the theme in the context of his enquiry into "sense of direction" and disorientation: "The experience of 'being turned around' results when sensory impressions are registered which cannot readily be fitted into the general scheme or which belong to a context different from the one in control" (Lund, 1930, p. 61): from the momentary confusion associated with this process of reorganization of information, a "new mental set" emerges, making subjects feel that they are once more "turned right". Far more recently, the feeling of being "turned around" has been discussed by Erik Jonsson, who provides us with an account of his first "strange and … unpleasant experience" (Jonsson, 2002, p. 15) of this kind: on arriving in Cologne in 1948, he was confused by perceiving "*Sunrise in the west!*" (ibid., p. 13) above the steamboats on the Rhine. Once he had embarked on his river cruise, this apparently "unresolvable [spatial] conflict" abated, only to reappear when he got back to the city that evening: "A very strange and most unpleasant experience. Afterwards the boat was going south, the Rhine flowed in the wrong direction, and the sunset was in the east. I left Cologne on the next train" (ibid., p. 15). As earlier noted, he was to relive this same experience of being turned around some years later in Paris.

"Boys, these are our own tracks, and we've actually been circussing round and round in a circle for more than two hours, out here in this blind desert!" (Mark Twain, *Roughing It,* 1872, p. 695). From Mark Twain, to Tintin (*Tintin au pays de l'or noir, On a marché sur la lune*), to Tolstoy (*Master and Man*, 1895), to Charles Sealsfield (*The Cabin Book or Sketches of Life in Texas*, 1844), the popular belief that people who are lost walk round in circles and end up retracing their own steps is a recurrent theme in literature, film, and TV fiction. In Arthur Upfield's crime novel *Man of Two Tribes*, Detective Napoleon Bonaparte of the *Queensland Police Force* describes an episode of "walking straight in circles" in the Australian bush during a missing person hunt, when a certain Maddoch attempted to return on foot to the caverns from which he had recently come:

> He would start walking to the north, […] opposite to the line of march. […]. Within minutes he would veer to the left or right, according to which leg was longer than the other. There being no natural objects to guide him, to lure him, he would inevitably walk in a circle.
>
> (Upfield, 1956, p. 170)

The phenomenon of spiralling in misorientation has been documented by a number of scientists, since first reported in the earlier cited debate on instinct among Darwin and his contemporaries on the pages of *Nature*. The tendency to engage in this kind of circular movement has been investigated in both animals and humans: the zoologist Asa Schaeffer maintained that all living creatures, from amoeba to humans, tend to move in circles (Schaeffer, 1928). In animals, spiralling is often interpreted as an intentional behaviour, fulfilling the aim of not straying too far from the den when in flight from a predator. In humans, Viguier claimed that the observed pattern of deviation from a straight course almost always tends towards the left, arguing that this is most plausibly due to the fact that the right leg is typically longer than the left (Viguier, 1882, p. 19).

For Schaeffer, on the other hand, the fact that blind persons engage in spiralling independently of whether walking, swimming, or driving a car, implied that differences in the length of their legs cannot be a causal factor: "Persons who lose their way in forests, snow-storms, fogs, etc., go 'round in circles', because in those persons the orienting senses are not functioning, and then the spiralling mechanism guides the path" (Schaeffer, 1928, p. 394). Lund, on his part, after studying the movement of students who had been blindfolded and asked to cross a flat open space in a straight line, found that 55 per cent of them veered to the right, and 45 per cent to the left after about 30 metres, leading him to revive the hypothesis of asymmetries in leg size (Lund, 1930). In our own day, while the folk notion of wandering around in circles remains popular, on the scientific front there is still uncertainty concerning both the true incidence of the phenomenon and how to interpret it. Recently (2009), the German scholar Jan L. Souman and his research team have revisited the topic, conducting field experiments with subjects walking in forest settings and the Sahara desert or crossing flat surfaces. They claim to have obtained the first truly empirical evidence that human beings tend to go around in circles when attempting to walk across unknown territory in the absence of landmarks. When it was dark or when they were blindfolded, their study participants generally continued to walk in circles, while if they could see the sun they managed to steer a straight course, even when hampered by environmental obstacles, such as forest vegetation. Souman and colleagues do not attribute this movement to differences in leg length, but to the involvement of different sensory channels, the motor and cognitive functions: "This drift in the subjective straight ahead may be the result of accumulative noise in all components of the sensorimotor system" (Souman et al., 2009, p. 1541). They conclude that: "Ironically, in the age of ubiquitous navigation systems in airplanes, cars, and even mobile phones, we are only beginning to understand how humans navigate through their environment, exploring uncharted terrain" (ibid.).

Maps and routes

Let us go back to where we left our tourist. He now needs to locate his destination, the museum, on the map by interpreting common cartographic symbols for streets, watercourses, monuments, etcetera, and drawing the right conclusions about the direction to take and the route to follow, an operation whose complexity and relative likelihood of error will depend on the number of turnings involved and the tourist's prior experience.[6] In the geographer's terms, as we have seen, a (geographical) map is a scaled-down, approximate, and symbolic representation of a given portion of the earth's surface. From the cognitive psychology perspective of O'Keefe and Nadel: "A map is the representation of a set of connected places which are systematically related to each other by a group of spatial transformation rules" (O'Keefe and Nadel, 1978, p. 86). The convention of orienting the map towards north may be of help to the tourist if he is familiar with the cardinal points. The real issue with locating his destination depends on which way his body is facing in relation to the orientation of the map. In the words of O'Keefe and Nadel: "Reading maps is a more complicated affair than following route instructions. [...] First, it is necessary to locate one's place and one's orientation on the map. This can be done by identifying several surrounding landmarks which can be seen (or heard) from one's position, and looking for these on the map (or vice versa). Two such objects and their spatial relationship will place the observer on the map. Here we have a complicated process in which the objects and their relationships in egocentric space are translated into three objects in unitary, or absolute, space" (ibid., p. 87). If we find ourselves at a crossroads formed by four streets at right angles to one another, we can go left or right, continue straight on or turn back. At junctions with circular roundabouts, the number of options increases: for example, if we are using a GPS device, to go straight the recorded voice may direct us to take the "second right" off the roundabout, potentially disorienting us even further in interpreting our perceptions of space. In England, Swindon's *Magic Roundabout*, which is named after a well-known children's TV series, is popularly viewed as one of the worst intersections in the UK if not in the world, and in December 2007 was voted one of the "Scariest Junctions in the United Kingdom" in a *BBC News* survey. The *Magic Roundabout*, built in 1972 to regulate traffic converging from five different directions, is formed by an outer loop and five mini-roundabouts on which the traffic flows clockwise, arranged around an inner loop carrying traffic in a counterclockwise direction;[7] despite the disorienting effect of this complex design, the slow pace of the traffic appears to ensure that the roundabout enjoys an excellent safety record.

This still leaves our tourist with the challenge of identifying his precise position on the crossroads (or street or square). To do so, he crucially relies, first on being able to interpret the map's symbolic language, which may be more or less intelligible, and second on the scale of the map being large enough to inform him of his precise whereabouts, for example what side of the street, intersection or square he is currently on. If street names are marked on the map, this will clearly make it easier for him to work out his location, provided he is able to identify the name of at least one side street in the vicinity. A lack of precise symbolic markers, as when street names are not provided on the map or in the tourist's real-life surroundings, can make his task far more difficult: a common situation in many Japanese cities, especially Tokyo. The layout of a city can also contribute to heightening our sense of disorientation, as eloquently argued by Roland Barthes (see Chapter 8): "Quadrangular, reticulated cities (Los Angeles, for instance) are said to produce a profound uneasiness: they offend our synesthetic sentiment of the City, which requires that any urban space have a center to go to, to return from, a complete site to dream of and in relation to which to advance or retreat; in a word to invent oneself" (Barthes, 1982, p. 30). While Roland Barthes (see § 8.1) complains that the quadrangular form of the typical American city is disquieting and synesthetically unappealing, Yi-Fu Tuan is of a totally different view, finding himself on the contrary far more at ease with the regular and rational grid pattern of New World cities:

> This fear of losing my way has strongly affected my environmental preferences. Unlike many people, I prefer American towns with their geometric street patterns to Old World towns with their maze of narrow alleys. Old World towns are unfriendly to strangers, who must stay a long time to feel spatially comfortable. By contrast, the open grid characteristic of many American towns says right away, "Welcome, stranger." I dislike the tropical rainforest for the same reason: unless one is a native, it invites disorientation. I like the desert because it is an open map, with the sun serving as a dependable marker of east and west, and with sharply-etched landforms – visible from miles away – that unmistakably tell the visitor where he or she is.
>
> (2009, p. 94)

Alternatively to consulting the map, our disoriented tourist could ask a passer-by how to get to the museum, thereby obtaining an elaborate set of instructions or the specification of a *route*[8] for reaching his destination. In the terms of O'Keefe and Nadel: "Routes *imply goals which imply motivations*" (O'Keefe and Nadel, 1978, p. 83). Routes allow us to construct a grid of reference points in relation to which we can establish our own position, as well as where to turn within an egocentrically defined space. The directions provided must be followed in the given order and do not allow for error: "Routes, then, are rapid and easy to use, but because of their inflexibility very vulnerable and easily rendered inoperative by changes in the route statement or the environment" (ibid., 1978, p. 85). Disorientation ensues when the instructions provided are incorrectly interpreted or applied or the subject is unable to follow them:

> The major drawback of routes, their inflexibility, leads to another vulnerability. Consider the following ways in which one can get lost. The route instruction may be physically or mentally degraded (smudged word, forgotten orientation), landmarks or other guides may be destroyed or

> changed (snow, fallen trees), the translation of the instruction into behaviour may go awry (an elm is mistaken for a beech), a momentary lapse of attention leads to a subgoal being missed. It will be instructive to consider the strategies available when one loses the way or is forced to detour while following a route, because it is here that the differences between routes and maps are most clearly revealed. Getting lost is such a problem that routes often attempt to foresee possible lapses of attention or deviations and provide appropriate cautions, "the old track ... is now elusive", or corrections, "if you strike the lower road, simply turn right".
>
> (ibid., 1978, p. 84)

Choosing which strategy to adopt in suggesting a route is itself culturally situated. The specification of routes depends, among other variables, on the physical shape of the environment (upstream, downstream), on the presence of recognized landmarks (monuments), and on whether it is customary to draw on allocentric reference systems such as the cardinal points (one block to the north), typically more familiar to the habitués of cities with a grid layout, as for example in the US, than to the residents of European medieval towns in which the streets radiate out from the centre. In other cultural contexts, such as Japan, it is common to explain a route by sketching a map on any scrap of paper that comes to hand. If in following a route provided to us, we fail to implement the instructions in the right order, we may get lost and find it difficult to backtrack to a point at which we are sure of our bearings:

> Note that in all cases it is important to find not just any mountain or any stream, but to find the same mountain or stream. [...] An additional problem with an extended guide is once having located it to follow it in the correct direction. Orientations are even more difficult to re-establish once they have been lost. It is not enough merely to re-adopt the original orientation, since this was only appropriate along the original axis. Once off that axis an entirely new orientation is required. [...] Thus, orientations are only useful when both the position of the traveller (the body axis) and some external axis are specified; they are purely egocentric tools. If one is forced off one's route by an obstacle, rather sophisticated geometric calculations would be required to regain the original orientation and axis. Landmarks are also useless since, again, they mark the orientation only in conjunction with the original axis. The farther away the landmark, the more the goal will be missed. One of the major problems with an orientation is simply knowing whether one is lost or not.
>
> (ibid., p. 84)

It is beyond the scope of this book to describe the mechanisms underpinning the functioning of the brain and nervous system[9] with regard to orientation. For our purposes here, it is sufficient to recall that the human capacity to organize reality in terms of a mental map relies on the activity of a brain region known as the hippocampus,[10] situated in the medial temporal lobe. Dudchenko explains:

> The detail of the hippocampal cell organization is of interest from a cognitive mapping perspective, in that neurons [...] encode spatial information. It's likely that the hippocampus is part of a larger cortical-and subcortical spatial information processing stream that underlies the ability to recognize one's location, direction, and locomotor movements within a given space.
>
> (Dudchenko, 2010, p. 119)

Place cells, discovered in the 1970s, and other neural components connected with the hippocampus that play a role in our perception of space, are all part of a complex system of functioning within the brain, many aspects of which we have yet to truly unravel and comprehend. Dudchenko concludes: "The hippocampus is a component of an interconnected neural circuit whose function may be to keep us from getting lost" (ibid., p. 189).

The hippocampus as a cognitive map

What we have outlined so far about the workings of orientation may strike us as a complex and somewhat inaccessible description of an empirical process, yet the everyday case of the tourist who is trying to locate a museum in an unfamiliar city presents just this level of complexity. Furthermore, it is only one of a large number of possible scenarios of orientation/disorientation; alternatively, we could have examined the case of a person who becomes disoriented during a mountain walk, is not equipped with a map and has no particular goal beyond getting back to where (s)he started out from, or we might have explored what is involved in moving around in the dark, or navigating the open sea in a boat, etcetera. All of these situations share the key elements analyzed above: space, the body, the mind, other objects present in space and the relations among them. As we have seen, an orientation/disorientation process involves numerous variables and competences, from kinaesthetic to spatio-temporal data, to the dynamics of movement and how it is represented and perceived. A range of scientific disciplines (geography, cartography, psychology, neurology, computer science, communication studies, urban studies, architecture, philosophy, etc.) have worked to bring this analytical process to light, identify its key components, and find appropriate terminology for defining such components and the relations among them. The data involved is not supplied exclusively by the sense organs but also by numerous individual variables (cultural, social, psychological, moral, and affective).

Indeed, the orientation/disorientation process, by its nature, cannot involve subjects independently of their mental faculties, reduced to bodies situated in a given space and intending to move to another given point (place) in space. Thus Whorf maintains that: "Newtonian space, time and matter are no intuitions. They are recepts from culture and language" (Whorf, 1956, p. 153). From a neurocognitive perspective, Dudchenko on his part raises the possibility that moods, emotions, and spatial representations may be functionally connected: "From a wayfinding perspective, these findings may imply that imagining one's destination is an important part of getting from A to B without becoming lost" (Dudchenko, 2010, p. 153). Other studies (Hassabis et al., 2007) have shown that patients whose hippocampus has been removed have difficulty imagining future experiences: "This raises the intriguing possibility that imagination and memory are functionally related… Place cells… encode not only current locations but also intended destinations" (ibid.). Furthermore, states of fear and anxiety, as reported by Moita et al. (2004), can condition how an environment is represented through the place cells (ibid., p. 166).

In geography, as we have seen, the notion of space is the non-negotiable point of departure from which to approach the theme of orientation/disorientation. Indeed, space appears to be the point of departure in all the situations that we have evoked so far; we cannot even image a body or an object without space.[11] But what exactly is space? This is a key question: is space a prerequisite for conceptualizing the body, and the body the necessary condition for conceptualizing space? Is it possible to imagine a space without objects (bodies, things)? Is an empty space equivalent to an empty container, such as for example an empty room or an empty box? On contemplating space, we are confronted with profound questions, which have been debated for centuries by philosophers and scientists and which concern us here because they are highly pertinent to the theme of orientation/disorientation: is it possible to orient oneself in an empty space, real or imagined? Earlier in the chapter we referred to O'Keefe and Nadel's key work, *The Hippocampus as a Cognitive Map*. On its publication in 1978, this book marked a turning point in cognitive neuroscience and the enquiry into the neural bases of cognition, which had been sparked off by the discovery of place cells in the 1970s. The authors' basic argument was that dysfunction or lesions of the hippocampus are responsible for the loss of those functions that depend on cognitive maps. Dudchenko sums up their proposition as follows: "The hippocampal local system represents 'the spatio-temporal context within which […] knowledge is acquired' […]. Loss of hippocampus would impair this type of memory, and impair exploration and place learning" (Dudchenko, 2010, p. 128). Since the 1970s, extraordinary advances in information technology and the capacity to process huge volumes of data have underpinned both the discovery of other space cells and significant progress in our understanding of neural circuits in the region of the hippocampus. In 2014, John O'Keefe, Edvard Moser, and May-Britt Moser were awarded the Nobel Prize for Physiology and Medicine "for their discoveries of cells that constitute a positioning system in the brain", essentially a sort of "inner global positioning system or GPS", and for their contribution to understanding spatial disorientation in patients affected by dementia and Alzheimer's disease.

Notes

1 The title of the section is taken from the poem *One Art*, 1979, by Elisabeth Bishop.

2 "In the 1960s, behavioral geographers, environmental psychologists, and urban designers attempted to forge a new vision of human–environment relations firmly grounded in principles of human cognition and behavior. […] As a consequence of this energetic and stimulating movement, geography and psychology had joint intellectual custody of the terms *cognitive map, spatial cognition,* and *spatial behavior* from the 1970s onward. It was geographers who invested heavily in nurturing intellectual interest in human spatial cognition in the final decade of the 20th century" (Allen, 2004, Preface, p. XV).

3 See also McCarter and Schmidt (1998).

4 "The gold-on-black decals are 24 inches in diameter, larger than a large pizza but smaller than a manhole cover. They carry two kinds of information: directions for north, east, south and west, and the names of the nearest streets" (Dudchenko, 2010, p. 66).

5 "The current New York subway system was formed in 1940 when the IRT (Interborough Rapid Transit), the BMT (Brooklyn-Manhattan Transit) and the IND (Independent) lines were merged. […] The untenable mess of overlapping sign systems finally got attention in 1957 when George Salomon, typographic designer at Appleton, Parsons & Co., made an unsolicited proposal to the New York City Transit Authority (NYCTA) entitled 'Out of the Labyrinth: A plea and a plan for improved passenger information in the New York subways.' The unpublished typescript anticipated many of the suggestions for overhauling the signage of the subway system that Unimark would make a decade later." Shaw (2008).

6 See also Scharfe (1998).

7 Worst 10 Junctions in the World; http://www.autoexpress.co.uk/car-news/29798/world-s-worst-junctions#ixzz3TiWPxVQ8; http://www.autoexpress.co.uk/gallery/4813/world-s-worst-junctions-pictures#ixzz3TiW7Teq1. Accessed 2 April 2017.

8 O'Keefe and Nadel (1978) divide routes into two categories: "Careful examination shows that within each S-R-S instruction the emphasis can be on the stimulus or cue component, on the one hand, or on the response requirement on the other. We shall call the former type of instruction a guidance and the latter type an orientation or direction" (p. 82).

9 See also Damasio (2012).

10 For an introduction to the functioning of the hippocampus and the role of place cells, head direction cells, grid cells, and border cells, see Dudchenko (2010).

11 The sense of disorientation in other dimensions of space is brilliantly described in Abbott's *Flatland*, when the Square meets the Sphere: "An unspeakable horror seized me. There was a darkness: then a dizzy, sickening sensation of sight that was not like seeing: I saw a Line that was no Line. Space that was no Space. I was myself, and not myself. When I could find voice, I shrieked aloud in agony. 'Either this is madness or it is Hell.' 'It is neither,' calmly replied the voice of the Sphere, 'it is Knowledge; it is Three Dimensions: open your eye once again and try to look steadily'" (Abbott, 1884, p. 155).

3 Philosophies of disorientation

Der Weg des Geistes ist der Umweg (*The path of Spirit is a roundabout path*)
(Hegel)

On space

It is not my aim here to evaluate the neuroscientific contents put forward by O'Keefe and Nadel (1978), whose main objectives were to analyze the anatomy and functioning of the hippocampus and place cells on the basis of laboratory experiments with rats, and review the existing scientific literature on lesions of the hippocampus. However, from the perspective of cultural geography, it is of interest to note that, in their introduction O'Keefe and Nadel felt the need to devote almost 60 pages to the age-old debate about what space actually is, and particularly to philosophers' frequently conflicting positions on this theme, from the late 1600s up to and beyond Kant. They first noted that the controversy on space has largely revolved around dichotomous propositions such as psychological space vs. physical space, relative space vs. absolute space, and finally, innate space vs. learned space. The first of these pairs opposes perceptions of space as an operation of the mind, on the one hand, and views of space as an entity that exists independently of the existence of animate beings, on the other. O'Keefe and Nadel mainly focused on psychological space, which historically has been influenced by ideas about physical space and, according to some psychological theories, "is learned from or constructed on the basis of experience with the physical universe" (O'Keefe and Nadel, 1978, p. 7).

The distinction between absolute and relative space may be applied to both psychological and physical space. Absolute space is understood as a container that exists independently of objects but within which objects may be situated. Objects – including the observer – may move around the container without altering its basic structural framework. Relative space, in contrast, is defined by the relationships among individual objects or sets of objects and, ultimately, cannot be imagined without the objects that occupy it. The third dichotomous pair concerns psychological space only: "innate or *a priori*" implies that "the structure and function of those parts of the brain involved in generating space" are "specified by information in the organism's genes". The alternative to this biogenetic model is to view the structure and functions of the brain as the outcome of a living organism's specific experience.

Last but not least, a further question that arises within the overall framework of the debate about space, whether understood as a physical or psychological concept, is whether it is best expressed by Euclidean or non-Euclidean geometry. O'Keefe and Nadel proposed that in relation to psychological space – the conceptualization of space most useful to theories of cognitive neuroscience – "the metric of the cognitive map is Euclidean" (ibid., p. 10), stating that this methodological choice would ultimately be borne out or contradicted by data from studies of animal behaviour and research on the physiological properties of the hippocampus. In this context, O'Keefe and Nadel noted that they had based their own research on a suitably modified version of Kant's theory of psychological space, according to which psychological space is absolute and *a priori*. Dudchenko summarizes their conclusions as follows:

> A unitary representation of absolute space exists and is found in the brain, and [...] is not the product of experience or derived from relative space. It underlies our ability to perceive the world "coherently". This locale space corresponds to Kant's idea of absolute space, and the locale system in the brain encodes an allocentric representation of the world. In contrast, egocentric space is referred to as taxon space, and our sensory and motor systems work in this realm.
>
> (Dudchenko, 2010, p. 126)

Kant's position in this regard was decidedly at odds with the most authoritative views of his day – namely those of Berkeley, Newton, and Leibniz – who had divergent views on physical space, but concurred that psychological space was relative and based on experience (cfr. O'Keefe and Nadel, 1978, p. 10).

In the well-known correspondence between Samuel Clarke, a pupil of Newton's, and Leibniz (published in English, French, and German over the period 1717–1720), Clarke defended Newton's concept of absolute space, while Leibniz maintained that space could only ever be defined as relative. The aspect of this exchange that is most salient to our analysis here is Newton's conviction that absolute space was not directly accessible to the senses. The differences between Newton's and Leibniz's accounts of psychological space rested on two different ways of conceptualizing the elementary particles of the universe: Newton held the universe to be composed of matter made up of atoms, while Leibniz's view – more philosophical than scientific – was that monads were the ultimate indivisible unit. In his correspondence with Clarke, Leibniz attempted to show that psychological space was given by the relationships among monads as experienced from "the point of view of the 'percipient'" (O'Keefe and Nadel, 1978, p. 13), thus adopting a phenomenological position whereby a spatio-temporal container was not required to define such interrelationships: “I will show here, how men come to form to themselves the notion of space. They consider that many things exist at once and they observe in them a certain order of coexistence, [...]. This order is their *situation* or *distance*” (*Leibniz’s Fifth Paper*, cited in O’Keefe and Nadel, 1978, p. 14; emphasis in original). In other words, for Leibniz, it was impossible to conceive of the existence of space without objects.

Berkeley too rejected the concept of absolute space but on different theoretical grounds. He saw space as being relative in that it was linked to tactile experience:

> When I excite a motion in some part of my body, if it be free or without resistance, I say there is space; but if I find a resistance, then I say there is a body; and in proportion as the resistance to motion is lesser or greater, I say the space is more or less pure. So that when I speak of pure or empty space, it is not to be supposed that the word “space” stands for an idea distinct from or conceivable without body or motion [...].
>
> (Human Knowledge, cited in O’Keefe and Nadel, 1978, p. 15)

In Berkeley’s view, relative space was therefore based on sensations that were translated into concepts or ideas in the brain (“images of these sensations”). His form of empiricism, which is also known as “sensualism”, allowed for different types of relative space, connected to one another via tactile experience. The ultimate conclusion drawn by Berkeley was that nothing exists independently of our minds.

This preamble helps us to see that Kant, as argued by O’Keefe and Nadel, “combined these strong points [those of each of the theories] into a single powerful theory of space” (ibid., p. 16). Berkeley and Leibniz emphasized “the primacy of the concept of body or sensation over that of space, while [Newton] did not” (ibid., p. 17). Kant came to concern himself with these divergent notions of space, because – aside from a strictly geographical interest – he was struck by the problematic nature of Newton’s attempt to reconcile the notion of absolute space with that of relative psychological space. Newton argued that absolute space underpinned the laws of physics, while relative space was generated by human beings in their interaction with material objects. However, Kant felt that the inductive reasoning posited by Newton in reaching this conclusion was not a strong enough basis to justify the concept of infinite space. Furthermore, he was drawn to Berkeley’s and David Hume’s suggestion that a conscious mind could not have direct access to the physical world. The solution ultimately put forward by Kant was to recognize absolute space not as a property of the physical world, but solely as an innate organizing principle of the mind. In other words, “Space was a *way* of perceiving, not a thing to be perceived” (ibid., p. 19). Initially however Kant’s basic vision remained close to Newton’s idea of absolute space as a property of the physical world. A vision borne out by objects that are similar but incongruent, such as right and left hands, which may only be defined as occupying a left or right position in relation to the space containing them: "If there were only one hand in the universe, it would be impossible to say whether it was a left or right hand without recourse to an absolute spatial framework" (ibid., p. 20).

Kant's ultimate shift away from Newton's interpretation of absolute space in favour of a psychological interpretation was influenced by the thinking of one of his contemporaries, the Swiss mathematician and physicist Leonhard Euler. According to Euler, space was exclusively psychological, but depended neither on sensations, nor on minds that process sensations. Kant's reading of Euler led him to develop his full-fledged theory of knowledge (*Erkenntnis*), in which psychological space played a key role: "In addition to knowledge which originated from the external world and could therefore be called *empirical*, or *a posteriori*, Kant identified a category of *a priori* knowledge, defining it as 'knowledge that is [...] independent of experience and even of all impressions of the senses' (*Critique*, p. 42)" (ibid.). To the previously well-established classification of propositions into *a priori* (i.e., independent of experience and, in this sense, innate rules that may be verified analytically) and the *a posteriori* sort (empirical, and therefore verifiable synthetically), Kant added a third category of proposition, the *synthetic a priori*. Propositions of this type do not have a logical structure like the *analytic a priori* kind but neither may they be verified empirically like the *synthetic a posteriori* variety. They may only be verified via mental processes, as for example in the case of the geometrical axiom stating that "a straight line is the shortest distance between two points".

According to these highly complex developments in Kant's thinking, of which only the outline may be sketched here, the human mind uses two different faculties to perceive the external-material world: the faculty of sensibility which generates representations (termed intuitions by Kant) of events and objects in their spatio-temporal context and that of understanding, which classifies objects in terms of abstract concepts. A red ball, for example, would be represented by sensibility as a specific ball near a blue chair, whereas understanding would place it in the abstract categories of redness, balls, etc. Within sensibility, Kant theorized two different types of representation: empirical intuitions (e.g., impenetrability, hardness, colour) or the "matter" of representations of the unknowable external world, and pure intuitions based on the structure of sensibility itself (e.g., extension), which provide our representations with "form". The key implication of this way of classifying the processes involved in perceiving the outer world and representing it in the mind, is that space is a pure intuition of the faculty of sensibility. Kant did not provide an account of how we might come to have such *synthetic a priori* intuitions, which remain controversial to the present day. Similarly to language, they may either be acquired *ex novo* by each individual or depend on an "*a priori*" biogenetic predisposition (as argued by Noam Chomsky, 2016).

With regard to the concept of space in particular, explanations of how we come to acquire it range from cultural relativism (there is not just one unavoidable way of conceiving space), to a pure empiricist model (our view of space is determined by the fact that we live in and experience Euclidean space), to the idea that the human mind is pre-structured in such a way as to allow us to develop an individual concept of space based on experience.

At this point in their review of the key philosophical debates of the past and the present, O'Keefe and Nadel introduced their own neo-Kantian model. They assumed that evolutionary processes have led to the development of an innate mode of perceiving space, conferring subjects with accurate knowledge of aspects of external reality. Thus, their two key propositions are that three-dimensional Euclidean space is imposed on experience by the mind, and that this unitary framework – which underpins our perception of space as all-embracing and continuous – is a prerequisite to experiencing objects and their movements: yet "none of this work can provide a clear answer to the question of whether or not an innate spatial framework specifying metric three-dimensional space exists" (O'Keefe and Nadel, 1978, p. 152).

(Dis)orientation and Kant

Kant's thinking leads us to the heart of this chapter, in which we examine how the question of orientation/disorientation has been treated within the philosophical debate about space. In his own work on space, Kant, as both geographer and philosopher, contributed to the shift from viewing orientation as an empirical geographical concern to analyzing it as a mental process, using disorientation itself as a metaphor for this shift, a change in emphasis that is particularly clear in his 1786 essay, *What Does It Mean To Orient Oneself in Thinking?* (*Was heißt sich im Denken orientiren?*).[1] In that work, he drew a provisional balance in relation to the so-called *Spinozastreit* between Moses Mendelssohn and Friedrich Heinrich Jacobi, which has been viewed as the origin of "the philosophical concept of orientation" (v. § 4).[2] Kant's idea of synthetic *a priori* knowledge, still today at the centre of the philosophical and epistemological debate, was presented systematically in his 1781 key work *The Critque of Pure Reason*,[3] but had already appeared in his early writings on orientation. These early Kantian works on space are worth examining here, because in practice they gave rise to the discussion of notions of orientation/disorientation within philosophy. The examples of disorientation chosen by Kant (the right and left hands, the darkened room, the night sky in which the positions of the stars have been inverted) suggest why it is that we might become disorientated in the "dark night" of thought. It is not surprising that Kant was the scholar who prompted the progression from the empirical-geographical meaning of the term "orientation" to its metaphorical use in philosophy (and by extension in everyday language). Farinelli explains: "Kant wrote about philosophy but he was a teacher of geography. He himself identified the decisive turning point in his thinking as a shift from the empirical geography of the visible, to 'geography as reason' as he called *The Critique of Pure Reason*, or the geography of 'the dark space of our intellect' [...]. This shift involved adding the immaterial nature of the object under investigation to the darkness, or invisibility, of the mind, which is a chthonic phenomenon" (Farinelli, 2003, p. 58).[4]

For Kant as a geography teacher, the true meaning of the word orientation was the geographical one: "literally, to find the *sunrise*" (Kant, 1996, p. 8). Thus, on the one hand he strictly adhered to the original, etymological meaning of the term *orient*, as given by Descartes[5] and the French and German dictionaries of the seventeenth and eighteenth centuries. On the other hand, however, he never viewed geography lessons as solely consisting of the transmission of facts and data, but as having a broader educational mission. In the words of Louden: "A major portion of Kant's teaching was devoted to trying to enlighten his students more about the people and the world around them in order that they might live (pragmatically as well as morally) better lives" (Louden, 2011, p. 65).[6] Both Kant's *Physical Geography* and his last work, *Anthropology from a Pragmatic Point of View* (*Anthropologie in pragmatischer Sicht*), which he published privately in 1798, were based on his academic teachings, aimed in turn at enhancing his students' *Weltkenntniss.*[7] Elden comments: "This would usually translate as 'world knowledge,' but Wilson has suggested the felicitous 'cosmopolitan knowledge,' with 'cosmopolitan philosophy' for the related *Weltwissenschaften*. This knowledge of the world, for Kant, was integral to the moral and political life of the citizen" (Elden, 2011, p. 3).[8] Kant's practical and moral vision of his work as a teacher of physical geography, understood as an empirical science belonging to "that knowledge of the world 'which serves to procure the pragmatic for all the acquired sciences and aptitudes, [...] useful not only for school but also for life'" (Withers, 2011, p. 52),[9] is undoubtedly one of the reasons for which his essay on orientation should be viewed as the point of departure for the metaphorical extension of concept of orientation into the realm of philosophy. O'Neill observes that: "Philosophers and others have often used geographical images to set out their most basic question. The thought is quite intuitive: philosophical perplexity is like the experience of being lost, of not knowing the right way to go. This geographical picture of the starting point for philosophical inquiry is built into our language. When we are lost we stray or wander, hence we miss the right way, so may err morally, intellectually, or theologically" (O'Neill, 2011, p. 216).

Early in his geography teaching career, Kant produced two other short essays that are conceptually related to his essay on orientation: "Concerning the Ultimate Ground of the Differentiation of Directions in Space" (*Von dem ersten Grunde des Unterschiedes der Gegenden im Raum*) (1768) and "On Space" (*Von dem Raume*) (1770).[10] The notion of space therefore appears to be intrinsically bound up with that of orientation. At the same time, it is equally clear that scientific and philosophical enquiry on space, given its status as a phenomenon that is observed and experienced, predated the debate about orientation. In Kant's view, the aim of geography was to describe simultaneously occurring spatial relations which he conceptualized as the "substratum" of historical knowledge, in turn understood as an account of temporal relations.[11] Mendieta suggests that, for Kant, describing geographical space was fundamental for all geographical theorizing as well as for historical narratives: "Kant subordinates history to geography, for there is no history without a space [Raum] or place [Ort] in which it can take place. Even if history is a continuous succession of events [...], things in nature do not cease to change, and thus give rise to different geographies. Nonetheless, there is no history without geography, and the latter is the foundation of the former [...and] also of all other forms of geography" (Mendieta, 2011, p. 358). We therefore now examine in chronological order the three works by Kant which are particularly salient to this chapter, and which respectively addressed empirical space, practical and imagined orientation, and the metaphorical space of thought.

The first of these treatises, *Concerning the Ultimate Ground of the Differentiation of Directions in Space* of 1768, already contained Kant's basic definition of space as an a priori experience. For Kant, *a priori experience* is a hybrid intellectual notion, given that "absolute space is not an object of outer sensation", but rather a "fundamental concept" (Kant, 1992, pp. 371 *ff.*) which "first of all makes [it] possible" to determine position or relations amongst parts of matter, that is, among objects and bodies. As a fundamental concept, however, it cannot be dismissed as "a mere figment of the imagination" (*ein bloßes Gedankending*), despite the fact that it is intuitively perceived by the "inner sense", but difficult to pin down through reasoning. Identifying in the abstract an "accepted" or axiomatic concept (*angenommener Begriff*) that defines "the ultimate data of our cognition", always poses a challenge, but this is especially so "when an accepted concept has implications which contradict our most obvious experience" (Kant, 1992, p. 372).[12] Kant attempted to base his own arguments on concrete evidence, putting forward the example – which has been debated ever since by thinkers from Wittgenstein[13] to Heidegger and David Harvey, generally attracting more criticism than agreement – of the right and left hands of the human body. This example is of interest to us here for at least three reasons: first, on account of its methodological contribution to the scientific and philosophical debate on space (absolute versus relative space, see Newton, Hume, Berkeley, and Leibniz; Einstein and Wittgenstein), and especially to the methodological debate within geography (Harvey, Farinelli); second, because it is in line with the vision adopted in this book of the human body as the "origo" (Bühler) of all forms of deixis,[14] and third, because 18 years later Kant was to draw on it again in his 1786 treatise on orientation.

In his first work on the theme of space, Kant attempted to justify Newton's physical concept of absolute space, but at the same time he made concessions to Leibniz's mathematical method (see 2.1.7). The much cited – and frequently criticized – experiment of two dissimilar but complementary bodies, namely the right and left hands, which cannot be superimposed in three-dimensional space, was taken to prove that *right* and *left*, just as *above* and *below*, and *in front* and *behind* are key reference points for the human body. Kant concluded that the difference between the two sides must be caused by "the concept which we have of direction in general" (Kant, 1992, p. 367), which in turn necessarily points to the existence of an absolute space.[15] Wittgenstein – in treating the relativity of time with respect to measurement – detracted from the value of Kant's experiment as proof of the existence of absolute space, by pointing out that two lines of equal length in one-dimensional space may only be superimposed in two-dimensional space. In the same way, we might conceive of the right and left hands as perfectly congruent bodies which may be superimposed by transferring them to a space with a greater number of dimensions: "A right-hand glove could be put on the left hand, if it could be turned round in four-dimensional space" (Wittgenstein, 1922, TLP, § 6.3611).[16]

Kant and space

Only two years later, in Paragraph 15 of his Inaugural Dissertation "*On the Form and Principles of the Sensible and the Intelligible World*" of 1770 (1992), under the heading "On Space (Von dem Raume)", Kant was to develop and perfect his concept of space and the notion of *synthetic a priori* knowledge, based on the combination of *a priori* judgements and lived experience. Before examining the salient points of Kant's concept of space in relation to the notion of orientation, we should make a point of situating his treatise on space in its historical context; as outlined earlier in this chapter, contemporary enquiry was mainly focused on the concepts of absolute versus relative space. The debate surrounding this question, then as now, has often been viewed as nothing but implicit intertextuality and therefore only accessible to specialists in the field; however, in some cases, the cross-references have been clear, such as when Kant attributed the "English" with a "notion of space" "as an *absolute* and boundless *receptacle* of possible things" which belongs "to the world of fable" (*Welt der Fabeln*). Kant was alluding here to the thinking – touched upon earlier in the chapter – of Newton, Hume, and Berkeley, and contrasting them with Leibniz, who explicitly represented the opinion of "most of our people" [that is to say in Germany; M.S.] that space is "the relation *itself* which obtains between existing things, and which vanishes entirely when the things are taken away" (Kant, 1992, p. 397).[17] Kant rebutted both of these frameworks with the statement that: "*Space is not something objective and real*, nor is it a substance, nor an accident, nor a relation; it is, rather, subjective and ideal; it issues from the nature of the mind in accordance with a stable law as a scheme, for coordinating everything which is sensed externally"[18] (ibid., emphasis in original). The decisive novelty in Kant's proposed framework was his positing of ideal and subjective schemata that are somehow innate to the human capacity for knowing. David

Morris, whose work *The Sense of Space* is based on Merleau-Ponty's phenomenology of perception, provides the following explanation of Kant's use of the *schema* as a conceptual instrument: "Kant invokes transcendental schemata in order to explain how it is possible to subsume intuitions under pure concepts, how categories that are fixed *a priori* can be applied to changing *a posteriori* appearances" (Morris, 2004, p. 34).[19] Morris is referring here to *The Critique of Pure Reason*, written in the same period as Kant's work on orientation. Dünne has also commented on the term *schema* in relation to *The Critique of Pure Reason*, published in 1781, in which Kant used the expression *image* (*Bild*) instead of *schema*. Kant's concept of space as defined in his 1770 treatise has been described by Dünne and Günzel as: "The key turning point in the history of the concept of space"[20] (Dünne and Günzel, 2006, p. 30). Indeed, while in his first work on the theme (see above), Kant used the example of incongruent bodies in three-dimensional space as proof of the existence of absolute space, by the second work, space had become a "subjective archetype" (*subjektive Urbild*).[21] This did not imply understanding the concept of space as a process of individual perception, but as a "pure intuition" (*reine Anschauung*) (Kant, 1996, p. 396, emphasis in original), while space itself was now the "first *formal principle of the sensible world*" (ibid., p. 398, emphasis in original), that is to say, a preconceptual concept. Space could only be conceptualized in this way if "*apprehended* concretely" (ibid., p. 396, emphasis in original), similarly to the axioms of geometry.[22]

In *The Critique of Pure Reason*, Kant devoted a further 10 pages or so to space, in which he reaffirmed the concepts put forward in his treatise *On Space* of over a decade earlier and clarified them in relation to his new construct of *synthetic a priori* knowledge. It is beyond the scope of this chapter to analyze Kant's line of argument point by point. Rather, we focus now on two aspects that are of particular relevance to the question of orientation. First Kant used an example to illustrate his idea that: "Space is a necessary representation, *a priori,* which is the ground of outer intuitions. One can never represent that there is no space, although one can very well think that there are no objects to be encountered in it. It is therefore to be regarded as the condition of the possibility of appearances, not as a determination dependent on them, and is an *a priori* representation that necessarily grounds outer appearances" (Kant, 1998, p. 158).[23] In Morris' terms, "Kant's transcendental argument claims that spatial experience is constituted by an *a priori* cognitive structure" (Morris, 2004, p. 6). In other words, we are innately predisposed to experience space, a view explicitly confirmed by Kant himself: "Only this first formal ground, e.g., the possibility of a representation of space, is innate, not the spatial representation itself"[24] (Kant, 1973, p. 136).[25] Morris continues his analysis of Kant's concept of space by criticizing William James' theory of empirical reasoning,[26] on the grounds that: "What might now be called the neurobiology of the body is constitutive of spatial experience, as if experience were rooted in fixed structures of a body-machine" (Morris, 2004, p. 6). Morris adds: "Our sense of depth and space is rooted in living dynamics that inherently cross body and world; and our sense of depth and space expresses the sense of the crossing of body and world. Our sense of space is not constituted by cognitive or neurobiological structures that are merely on our side of things; our sense of space is enfolded in our outside, in a world that crosses our body"[27] (Morris, 2004, p. 6). Morris thus leads us to the second of two key aspects of Kant's chapter on space in *The Critique of Pure Reason*, in which he once more specifically addressed the subjective dimension of his concept of space. Without this subjective dimension, the very idea of space could not exist: "We can accordingly speak of space, extended beings, and so on, only from the human standpoint" (Kant, 1998, p. 157).[28]

This concept of space – and above all the introduction of the subjective dimension of perception – must necessarily be recognized as significant: "A turning point in the history of the concept of space".[29] Kant, after using the example of the right and left hands in his 1768 work (see § 2 above) to justify a position reconciling relational and absolute space whereby the former implies the latter, went on to innovatively introduce the perception of space as an innate human possibility of preconceptual experience in his treatise *On Space* two years later (see above, § 3). Kant reached this conclusion, as we have seen after following the correspondence between Leibniz and Clarke on the question of space, in which Leibniz defended a structural-topological and mathematical concept of space, while Clarke drew on the mechanistic cause-effect model of space theorized by Newton: Kant was therefore the first to propose the perception of space as a preconception, that is to say as a possibility of experience that is innate to the human being.[30] In his 1786 work on orientation, he would incorporate this *a priori* approach to space as a mode of perception into a lifeworld (*Lebenswelt*) philosophical model. The notion of *Lebenswelt*, first touched on by Kant, later featured strongly in the thinking of Edmund Husserl and was further developed in phenomenological and post-phenomenological philosophy (see § 3.6–3.7).

Throughout this succession of conceptual models, Kant's descriptive geography – as defined in *The Critique of Pure Reason* – occupied a key place in his hierarchical order of knowledge: God (theology), soul (psychology), and world (geography). The subject of knowledge was positioned at the heart of experience and reflection, just as in modern cosmology, the earth – viewed from the sun by a virtual observer – is seen as a dot among other dots, moving about space.[31] As we have seen, the issue of space is an age-old theme which continues to be debated today, in both physics and philosophy. It is beyond the scope of this book to discuss the debate on space within physics, despite its indirect bearing on the concerns of geography. This relationship with geography is particularly evident in the definition of space given by Albert Einstein: space is not a container which without the objects it contains would resemble an "empty box" (Newton); nor is it a purely relational entity that would not exist in the absence of the objects it contains (Leibniz). For Einstein space was "extended substance" (i.e., physical objects are not in space, but are spatially extended) in the Cartesian sense, not absolute space as understood by Newton, nor apriority[32] as proposed by Kant. However, Kant's framework does not entirely contradict the interaction of objects hypothesized by Einstein, given that the inclusion of perception in the philosophy of the living world situates the subject of experience in space, that is to say, in reality. Nonetheless, in Kant, this subject is philosophically isolated, and thinks transcendentally, while space is not simply given, nor does it simply exist, but is determined by the interaction of bodies in space or by human actions, as we learn from the social sciences that concern themselves with space, not least social geography.[33] In relation to concrete uses of gender (see Chapter 4) and the "conquest" of social space, as discussed by Löw, orientation is indeed a highly contemporary theme, at both the practical and theoretical levels.

What does it mean to orient oneself in thinking?

Kant's essay *What Does It Mean To Orient Oneself in Thinking?* (1996) dates to 1786, and therefore to the period between the first (1781) and second (1787) editions of *The Critique of Pure Reason.* The text was influenced by the context in which it was written and published, which I now briefly outline in order to flesh out the background to the conceptual shift from a geographical to a philosophical perspective on space and orientation. In what is known in the history of philosophy as the "Spinozism" or "Pantheism Dispute" (*Pantheismusstreit*),[34] Friedrich Heinrich Jacobi, a strongly Christian, and in this sense anti-Enlightenment, thinker, had accused Spinoza – viewed in "Germany as the spectre of a heretical Jew and atheist" (Stegmaier, 2008, p. 64) – of "philosophical theology" and therefore of atheism. According to Jacobi, appealing to philosophical reasoning on matters of religious revelation would end by destroying faith and laying the ground for nihilism. After the death of Gotthold Ephraim Lessing (1781), a leading exponent of the German literary Enlightenment, Jacobi wrote a letter to Moses Mendelssohn, founder of the Jewish Enlightenment, in which he accused Lessing of adhering to "Spinozism". Mendelssohn felt obliged to defend his friend Lessing.[35] He thus attempted to reconcile Jacobi's opposition between faith and reason by appealing to the concepts of common sense and speculative thought, offered as twin guiding lights enabling us to find the right path, in other words to orient ourselves. In Mendelssohn's terms, we do not require external points of reference such as the sunrise or a compass to orient ourselves, because reason is self-orienting in nature.[36]

Mendelssohn died in 1786. His sudden decease, allegedly brought on by the bitter ongoing dispute with Jacobi, sparked off intense debate among contemporary scholars, and thus Kant was obliged to join in the controversy. Mendelssohn himself had begun to make metaphorical use of the concept of orientation, for example by speaking of "being guided by *common sense*" in his "allegorical journeying" (*allegorische Wanderung*).[37] The philosophical metaphor was even clearer in one of Mendelssohn's letters to Lessing: "Each time that my speculation appears to stray too far from the royal road of common sense, I stop and try to recover my bearings. I look back to the point from which I started out and try to compare my two guides".[38] Kant esteemed Mendelssohn, although he did not share his philosophical position. He wished to defend him and at the same time, he had a personal interest in refuting Jacobi's use of passages from *The Critique of Pure Reason* to critique Spinoza. Furthermore, Kant saw intervening in the debate as an opportunity to uphold the validity of the *Critique*: this led him to develop "the key basis" for critical thinking about orientation.[39]

Kant's line of reasoning spanned three phases: in moving from geographical to mathematical, and subsequently from mathematical to logical orientation, he made two successive transitions: the first from the "real" and concrete meaning of the term to its abstract meaning, albeit remaining within the confines of a pregiven space, and the second to its purely metaphorical meaning. He described and defined geographical orientation in the following terms: "In the proper meaning of the word, to *orient* oneself means to use a given direction (when we divide the horizon into four of them) in order to find the others – literally, to find the *sunrise* [...] Now if I see the sun in the sky and know it is now midday, then I know how to find south, west, north, and east" (Kant, 1996, p. 8).[40] In other words, if we are aware of the rising of the sun and of the fact that it takes place to the east – a condition that is implicit in Kant's use of the verb *to know* – then at noon, we will be able to identify all four cardinal directions. However, this requires a certain type of reflection, or as Kant calls it, of "feeling" on our part as observing subjects. Specifically, we must have the capacity to distinguish between our right and left hands. As noted earlier, Kant had already used this example in his first work, not yet in relation to orientation but solely concerning how we conceive space (see § 2 above):

> I also need the feeling of a difference in my own subject, namely, the difference between my right and left hands. I call this a feeling because these two sides outwardly display no designatable difference in intuition. If I did not have this faculty of distinguishing, without the need of any difference in the objects, between moving from left to right and right to left and moving in the opposite direction and thereby determining a priori a difference in the position of the objects, then in describing a circle I would not know whether west was right or left of the southernmost point of the horizon, or whether I should complete the circle by moving north and east and thus back to south. Thus even with all the objective data of the sky, I orient myself geographically only through a subjective ground of differentiation.
>
> (Kant, 1996, pp. 7–8)

Kant therefore theorized a space that was given, and in this sense objective, but conceded that uncertainty would come into play when subjects needed to orient themselves in relation to a particular territory, city and so on. To illustrate this, he had earlier argued in the *Differentiation of Directions in*

Space (1768) that the difference between left and right is also critical to the reading of a text, given that not only do we distinguish between the top and bottom of a page, but also whether the characters run from left to right or vice versa. In relation to orientation, he went on to cite the imaginary example of the star-filled sky in which the positions of the constellations are inverted along the east–west axis: in such an eventuality "even the astronomer – if he pays attention only to what he sees and not at the same time to what he feels – would inevitably become *disoriented*" (ibid., p. 9).

The example of the astronomer led Kant to discuss the second form or method of orientation: "Now I can extend this geographical concept of the procedure of orienting oneself, and understand by it orienting oneself in any given space in general, hence orienting oneself merely *mathematically*" (ibid.). Just as in geographical orientation, here too Kant hypothesized a given space, though elevating the conceptual reach of human orientation to a higher level of abstraction, described as "merely mathematical." This level did not feature defined aids to orientation, such as the sunrise, the cardinal points or in general any kind of empirical framework. The subject of mathematical orientation no longer based reality on objects extended in different "directions" (*Weltgegenden*), as Kant had proposed in his 1768 work, but now constructed mathematical objects based on intuition, without reference to empirical experience.[41] Strictly speaking, Kant saw this higher orientation process as drawing on a formal language, stripped of semantic interpretations,[42] but even so – if we were to find ourselves in darkness, but in a room that we already knew[43] – the left–right distinction would remain anchored to the "position of the body in the world in which we orient ourselves" (Stegmaier, 2008, p. 85).[44]

This key role of the body as deictic "origo" (Karl Bühler, see footnote 21) was again illustrated by Kant using a practical example:

> In the dark I orient myself in a room that is familiar to me if I can take hold of even one single object whose position I remember. But it is plain that nothing helps me here except the faculty for determining position according to a subjective ground of differentiation: for I do not see at all the objects whose place I am to find; and if someone as a joke had moved all the objects around so that what was previously on the right was now on the left, I would be quite unable to find anything in a room whose walls were otherwise wholly identical.
>
> (Kant, 1996, p. 9, original italics)

Kant cited a second example from his everyday experience in support of the first: "But I can soon orient myself through the mere feeling of a difference between my two sides, the right and left. That is just what happens if I am to walk and take the correct turns on streets otherwise familiar to me when I cannot right now distinguish any of the houses" (ibid.). The crucial question here is if there is actually any difference between these last two examples (the darkened room and street) and the starry sky with the inverted constellations.[45] If we can find no difference, then we must ask ourselves what it is that distinguishes geographical from mathematical orientation.

Clearly, the difference lies in the fact that the astronomer observes "the objective data of the sky", that is to say, the stars, while the person in the darkened room, reaching out to touch items of furniture, cannot see anything, but only conserves "in memory", that is to say, in his or her imagination, the items of furniture and their positions. Ultimately, the very notion of mathematical orientation is based on that of absolute space as understood by Descartes and Newton, in that "the mathematical-geometric measures of length, width and depth may be calculated independently of the location of an object" (Quadflieg, 2009, p. 279). Thus "mathematical orientation" is to be viewed as a purely mental procedure. However, what geographical and mathematical orientation have in common is their reliance on the subject's body as a point of reference, on the "emotional" distinction between right and left and the aid to orientation supplied by sight (starry sky) or touch (darkened room).

If we now assume that logic is the meta-language of mathematics, it becomes clear how we may make the further leap from mathematical to logical orientation. If the transition from geographical to mathematical orientation involves shifting from that which we can see and geographically describe to that which we can mathematically calculate on the basis of given coordinates, the only logical follow-up is to substitute concrete space with metaphorical space, that is to say, with a space that is solely imagined and cannot be perceived by sight or touch. In Kant, this last step was marked by a move away from the real-life metaphor of the darkened room as well as by a change in terminology: "Finally, I can extend this concept even further, since it could be taken as consisting in the faculty of orienting myself not merely in space, i.e. mathematically, but in *thinking* in general, i.e. *logically*" (Kant, 1996, p. 9, original italics). He went on to elaborate:

> By analogy, one can easily guess that it will be *a concern of pure reason* to guide its use when it wants to leave familiar objects (of experience) behind, extending itself beyond all *the bounds of experience* and finding *no object of intuition* at all, but merely space for intuition; for then it is no longer in a position to bring its judgments under a determinate maxim according to objective grounds of cognition, but solely to bring its judgments under a determinate maxim *according to a subjective ground of differentiation* in the determination of its own faculty of judgment.
>
> (ibid., pp. 9–10, italics added)

Kant was thus now positing a subject no longer bound by the limits of empirical experience, whose self-orientation was not based on its encounter with concrete objects (such as the north star or the memorized location of a piece of furniture) but on a principle upheld by subjective reflection. The subjectivity of this theoretical-logical decision-making process was not to be viewed as mere arbitrariness, no more than was the left–right distinction, but rather as a form of ontogenetic conditioning, reinforced by practice.[46] The subject was thus – in line with Kant's own theory of knowledge[47] – a transcendental one: "This subjective means still remaining is nothing other than the reason's feeling of its own needs" (ibid., p. 10). In this last sentence, Kant introduced the specific signification of "feeling", previously invoked in the context of right–left orientation, as a "feeling of its own needs". Once again, we find a blending of the empirical subject with the transcendental subject in Kant's thinking given that: "Reason does not feel; it has insight into its lack and through the *drive for cognition* it effects the feeling of a need" (ibid., p. 12, fn. 1, original italics).[48] It is fully in keeping with the spirit of the Enlightenment that Kant should assign the work of philosophy, and therefore the ethical and moral orientation of reason, to a free and thinking subject. He had already advanced such a proposition in his famous essay *Answering the Question: What is Enlightenment?* (1784), but now made it even clearer and more explicit in "What Does it Mean to Orient Oneself in Thinking?" by drawing yet again on the metaphor of the darkened room to illustrate the role of reason in guiding us away from the darkness of obscurantism: "For *orienting* itself in thinking, solely through reason's own need, in that immeasurable space of the supersensible, which for us is filled with dark night" (ibid., original italics, p. 10).[49] In the final section of this chapter, let us briefly review how, and with what implications, the metaphor of dis-orientation was later taken up in phenomenological and post-phenomenological thought.

The metaphor of dis-orientation

By this stage in the development of his thinking, Kant had definitively raised orientation to the status of a philosophical issue, although he continued to display a preference for describing it in terms of geographical images. Such as when – in the spirit of David Hume, cited in *The Critique of Pure Reason* as one of the "geographers of human reason" (Malpas and Thiel, 2011, p. 197) – in presenting his philosophical topography of reason, he spoke of the "guide and compass of speculative thought in its incursions into the field of supernatural objects" (Stegmaier, 2008, p. 91), the "map of the country" of reason (ibid., p. 93), and the "land of truth" (Malpas and Thiel, 2011, p. 197); while in *Anthropology from a Pragmatic Point of View* (1798) we read once more that:

> The field of sensuous intuitions and sensations of which we are not conscious, even though we can undoubtedly conclude that we have them; that is, obscure representations in the human being (and thus also in animals), is immense. Clear representations, on the other hand, contain only infinitely few points of this field which lie open to consciousness; so that as it were only a few places on the vast *map* of our mind are *illuminated*.
>
> (Kant, 2006, p. 24)[50]

Kant, the geographer-philosopher, made rich use of metaphor in leading us from spatial orientation in the physical sense, to philosophical orientation in thinking. More generally, the representation of space not only brings a cultural perspective to bear on the theme of disorientation but also leads to the use of spatial metaphors in science and everyday language as a mental operation of mapping between two or more domains of conceptualization of experience (see Lakoff and Johnson, 2003).

Ivan Almeida opens his essay on the notion of orientation in Kant (Almeida, 2004) by citing Richard Rorty's "provocative aphorism": "It is picture rather than propositions, metaphors rather than statements, which determine most of our philosophical convictions" (Rorty, 1979, p. 12). He adds that: "If this is so, then geography could be viewed as one of the sources of philosophy, given the extent to which its concepts help to articulate the questions that the latter never ceases to raise" (Almeida, 2004, p. 1). Among the geographical concepts listed by Almeida, we encounter yet again that of the labyrinth (see Chapter 7). For Kant, in *The Critique of Pure Reason*, the path of knowledge became the journey towards an island surrounded by stormy seas:

> We have now not only travelled through the land of pure understanding, and carefully inspected each part of it, but we have also surveyed it, and determined the place for each thing in it. But this land is an island, and enclosed in unalterable boundaries by nature itself. It is the land of truth (a charming name), surrounded by a broad and stormy ocean, the true seat of illusion, where many a fog bank and rapidly melting iceberg pretend to be new lands and, ceaselessly deceiving with empty hopes the voyager looking around for new discoveries, entwine him in adventures from which he can never escape and yet also never bring to an end. But before we venture out on this sea, to search through all its breadth and become certain of whether there is anything to hope for in it, it will be useful first to cast yet another glance at the map of the land that we now leave, and to ask, first, whether we could not be satisfied with what it contains, or even must be satisfied with it out of necessity, if there is no other ground on which we could build.
>
> (Kant, 1998, p. 354)

The contemporary importance of Kantian metaphor in the orientation of scientific and philosophical thinking, as a means of bridging the gap between reality and theory, is reflected in a comment by Hans Blumenberg about the little-known work of the Swiss chemist Friedrich Miescher, who discovered nucleic acid in 1869.[51] Blumenberg observed that Miescher drew on mechanistic images to illustrate organic processes (the clock mechanism, vital energy, forces of tension), an "implicit" "underlying set of metaphors", "which served to orient him, as he stumbled his way into hitherto inaccessible regions" (Blumenberg, 1981, p. 411). So too, orientation and disorientation in human and non-human organisms may be explained using an anthropomorphic language: from such a perspective, defining the act of orienting oneself as intentionally searching for a goal in space is to be seen as a metaphor, while in practice orienting oneself is a *phenomenon* of moving about a real space, not to be exclusively interpreted as an intentional search but also as an instinctive form of a predetermined behaviour. At this point, *phenomenon* is the key word leading us on to discuss the use of the term dis-orientation in philosophical language. Kant (re)elaborated his geographical and philosophical reflections on space (1768 and 1770) and on (dis)orientation in thinking (1786) in the purely philosophical rather than geographical context of his *Critique of Pure Reason* (1781/1787). In the revised second edition of 1787 in particular, in an attempt to provide a unified account of reason and experience, Kant introduced the terms *Noumena* (objects perceived only at the level of our intellectual, though not yet scientific, understanding) and *Phaenomena* (objects perceived in terms of sensibility). This brings us to Kant's ideas on the distinction between subject and object in our perception of the *Lebenswelt* and, in a further step, to the methodological issues posed by a phenomenological approach to scientific research.

Lifeworld and phenomenology

Any treatment of orientation in the sense of a purely abstract operation of thought necessarily overlaps with that of orientation in the concrete, non-metaphorical sense, given that in both cases the process at issue is mental, the only difference being that the latter requires real spatial coordinates identifying a starting point, a displacement, and a destination. Today, the concept of *Lebenswelt* is associated with the thinking of Edmund Husserl and modern phenomenology, but as we have seen it was first introduced by Kant:

> Kant's procedure shows very clearly that it was he and not the twentieth-century phenomenologists who first conceived the value-laden objects of cultural experience as epistemological ultimates. It was he, not ourselves, who first discovered the irreducible connection of "fact" and "value" in the Lebenswelt [world of lived experience]. For Kant argues that the objects of cultural experience, no less than those of primary sensation, are constituted by the mental organization of the perceiver.
>
> (Hirsch, 1976, p. 101)

Since its introduction by Edmund Husserl at the beginning of the twentieth century, the philosophical and methodological approach of phenomenology – the science that, literally, concerned itself with phenomena, or appearances – has been the object of lively debate among its followers and critics. In any case, the key focus of phenomenology was and remains the question of how we experience the real world. In the geographical domain, phenomenology brought the subject back to the centre of discourse. The phenomenological method of description was applied by humanistic geographers in the 1970s and 1980s

(Relph, Buttimer, Seamon, Ley, and Tuan), with the aim of "reveal[ing] the 'subjective' experience of the agents of their life-world" (Gould and Olsson, 1982, p. 39). These geographers' exploration of the lifeworld was informed by the work of Husserl and later by that of Merleau-Ponty. In relation to space, which as we have seen is the key concept in the geographical analysis of disorientation, Merleau-Ponty distinguished between objective and existential space and emphasized the key role of the body in our perception of space: "not the body as a chunk of space or a bundle of functions but that body which is an intertwining of vision and movement" (Merleau-Ponty, 1964, p. 16). Humanistic geographers investigate the concepts of lived space, place and dwelling, thereby introducing a subjective dimension into spatial analysis. For Ann Buttimer, the "lifeworld" where "experience is constructed and where consciousness is revealed" (Buttimer, 1976, p. 280) encompasses both the physical and the social and we usually take it as given in the course of our daily lives. This view is shared by Relph, who claims that the lifeworld is "[t]hat world of ambiguities, commitments and meanings in which we are inextricably involved in our daily lives but which we take very much for granted" (Relph, 1976, p. 3), and by Seamon who defines "at-homeness" as "the usually unnoticed, taken-for-granted situation of being comfortable in [...] the everyday world" (Seamon, 1979, p. 70). Whether defined as "a return to the things themselves" (Husserl), "a way of seeing" (Heidegger), or "the essence of perception" (Merleau-Ponty), phenomenological approaches may be considered a "method" of accessing understanding of the world. In geography: "Rather than the aggregation of human beings into statistics, figures or mappings, phenomenological approaches were perceived to focus back onto the human as an individual figure; the medium and recipient of experiences, sensations and fleshy embodiment" (Lea, 2009, p. 372). The notion of *Placelessness*, inspired by Heidegger's concept of *dwelling*,[52] once again raised the issue of disorientation in terms of "the undermining of the importance of place for both individual and cultures, and the casual replacement of the diverse and significant places of the world with anonymous spaces and exchangeable environments" (Relph, 1976, p. 143). Thus from a phenomenological perspective, orientation and disorientation are crucial issues, related to our living experiences, to our bodies (Husserl's *Leib* or living body and Merleau-Ponty's *chair* or flesh) and emotions. Furthermore, according to Ahmed, phenomenology "makes 'orientation' central in the very argument that consciousness is always directed 'toward' an object, and given its emphasis on the lived experience of inhabiting a body, or what Edmund Husserl calls the living body (Leib)" (Ahmed, 2006, p. 2). She further argues that orientation is part of Husserl's theory of intentionality (the example of the table), one of the tenets of the phenomenological paradigm.[53]

The post-phenomenological approach

The term post-phenomenology was introduced by the philosopher Don Ihde (2003) to mark a departure from the framework of classical phenomenology and in particular from the thinking of Husserl, which assumed an undifferentiated body-subject. Ihde further maintains that post-phenomenology "finds a way to probe and analyse the role of technologies in social, personal, and cultural life" (Ihde, 2009, p. 23). However, post-phenomenological theories are somewhat fragmented and do not currently represent a single coherent school of thought. Generally speaking, they focus on new ways of seeing "our understanding of the relation of the body to the world" (Lea, 2009, p. 377), in light of knowledge gathered in the cognitive, biomedical and technological fields, as well as of "how particular technologies can mediate consciousness" (Ash and Simpson, 2014, p. 6). In geography, the post-phenomenological approach started out from a critical re-elaboration of the thinking of Husserl, informed by "Continental" post-structural theories (Bourdieu, Deleuze and Derrida), and continued with the development of non-representational theories (proposed by Nigel Thrift, Joyce Davidson and others), informed by the phenomenological ideas of Merleau-Ponty.

Post-phenomenological geographies contribute to the exploration of (dis) orientation by virtue of their "phenomenological concern with the embodied self in the world negotiated in relation to the fragmented and fragmenting subject" (Lea, 2009, p. 378). From a methodological point of view, post-phenomenology describes and classifies the various conditions (or *narrative frames* in the terms of cognitive science) under which the "human-as-flesh body" (Merleau-Ponty: *chair*) may be categorized in relation to the world, investigating phenomena of disorientation such as: "the intellectual experience of disorder, but [also] the vital experience of giddiness and nausea, which is the awareness of our contingency, and the horror with which it fills us" (Merleau-Ponty, 2006, p. 296). Such an approach acknowledges the role of emotion and affect, themes which have become increasingly central within the recent geographical debate: "Emotions are neither 'purely' mental nor 'purely physical', but ways of relating and interacting with the surrounding world. This relational account gives occasion for a double conception of emotional spatiality" (Simonsen, 2013, p. 17). The landscape and the place in which we live are reflected in our perception of them:

> This emphasises the need of individuals to reinforce their human selves as centres, in order to counter the depersonalizing tendencies that are incipient in every sensation. In allowing an understanding of what happens when the raw "power" of sensation overwhelms and threatens an individual's embodied sense of self, phenomenology offers opportunities to both describe and explain embodied geographies of health and illness.
>
> (Lea, 2009, p. 374)

Open and closed environments, bodies of water, forests, deserts, and mountains can all influence our moods. Joyce Davidson presents the lived experience of agoraphobia in the terms of Merleau-Ponty's phenomenology: "The subject of panic feels that the space they occupy is somehow qualitatively different from that of other people around them, they do not belong there and their only option is to run, usually homewards. Having 'lost the place' in this way, home often seems to be the imaginable space in which to rebuild your defences, away from the prying eyes and presence of others" (Davidson, 2000, pp. 652–653). Our perceptions of space become a key component of our wellbeing, while the loss of our usual points of reference can translate into extreme anxiety (or the uncanny feeling referred to in the Introduction, § 2). The emotional experience of getting lost can involve fear, even giving rise to panic attacks and stress:

> So long as lived space is successfully maintained, we have a sense of freedom to move as we please (within the limitations of our physicality), an easy relation with an environment that invites and encourage our involvement. But, if our sense of space breaks down we may find ourselves facing a repellent, unwieldy environment that restricts our freedom of movement, and threatens our sense of wellbeing.
>
> (Davidson, 2000, pp. 645–646)

In contrast, for Ahmed, when disorientated "we might even find joy and excitement in the horror" (Ahmed, 2006, p. 4). Willingness to get lost could be interpreted as a different and freer relationship with our body that enables: "shifts in one's sense and orientation in the world [...] so as to be transported fully into the magnificent chaos of life" (Braidotti, 2000).[54]

The aim of post-phenomenology is to advance our understanding of the nature of experience in relation to the bodies and sensations of human subjects, but viewing this experience as extending transhumanly:

> This allows the analytic focus to be decentered from and extended beyond the human body, taking in the inhuman (including, e.g., the agency of animals or objects) and the nonhuman (the pull of the world). Experience does not just occur within the body, confirming our boundaries and corporeality, but is a creative force distributed across bodies and the world.
>
> (Lea, 2009, p. 374)[55]

It also revisits the social construction of difference, an area that was not critically addressed by Merleau-Ponty. Thus, post-phenomenological geography conceptualizes the social as a "selective encounter between objects" (Ash and Simpson, 2014, p. 14), positing an "open and corporeal sense of difference" that underpins social, economic, and political categories (gender, race, age, and belonging): "The power of those categories comes from how the singularity of objects and their forces in the world are transmitted through and across populations of humans, but also non-humans" (ibid., p. 13). The emphasis here is no longer on the subject, but on the relations among a variety of objects – words, images, and bodies: "Post-phenomenology allows us to consider how objects have capacities for relation that humanistic forms of phenomenology would only assign to human beings" (ibid., p.12). Thus the study of the *lifeworld* comes to embrace the continuous circulation of human and non-human objects: "The social is then literally constructed by the objects that surround, pass over, miss and engulf human bodies in an often contingent manner, and which is not necessarily oriented around the concerns or projects of those involved in the social field" (ibid., p. 15). In this context, objects are defined as all reference to and, study of, relations with and among "non-human things", including "objects" such as "air or light as much as brains or computers, [...] kinds of memory [that] can be understood as sensual translations or partial comprehension of other objects retained by the objects that are involved in an encounter" (ibid., p. 14).

The reflection on objects[56] leads us to our next disorientation-related theme. If, on the one hand, orientation, as we have seen, draws on a complex set of components from the cognitive-intellectual, emotional, sensory-motor, cultural, and social domains, the post-phenomenological approach, on the other, allows us to overcome the separation between body and environment and to enter into the relations among the various human and non-human spheres, attributing them with meaning and situating them in a spatio-temporal dimension: "Body-subjects are visible-seers, tangible-touchers, audible-listeners, etc., enacting an ongoing interwining between the flesh of the body, the flesh of others and the flesh of the world" (Simonsen, 2013, pp. 16–17). As we have seen, in constructing the mental maps that structure our "awareness of our familiar environment", we may organize our images in different ways, by drawing for example on elements of the environment itself, myths, symbols, or abstract systems of reference. A mental map is the outcome of an encounter among a "variety of objects, such as words, image and bodies, that makes up the identifiable aspects of [a] culture" (Ash and Simpson, 2014, p. 14). The cognitive and environmental relations underpinning human and non-human spatial abilities and wayfinding strategies may be understood in light of Thrift's analysis of his attempt to learn a new language, Chinese, by repeatedly tracing the complex ideographic characters "in the air": "Words are remembered as gestures and so written into the body through movements as much as vision. The kinetic etching activates both hemispheres of the brain, and, in particular, those parts of it which involve motoric memory skills" (Thrift, 2009, p. 135). The process followed by Thrift in learning to write Chinese also takes place in orientation/disorientation and implies "the retention and transmission of memory between different objects, whereby particular shapes of line become translated into gestural movements in the limbs of the body and then translated into electrical signals in the brain and vice versa" (Ash and Simpson, 2014, p. 14).

Migration as a process of disorientation and reorientation

The experience of disorientation not only casts doubt on our location but also brings into question our very identity: "It is a way of describing the feelings that gather when we lose our sense of who it is that we are" (Ahmed, 2006, p. 20). Dis/orientation has to do with our position in space and time and, as we perceive it, is laden with identity-related, political, cultural, and sexual values: "Moments of disorientation turn our world upside down. Disorientation is a bodily feeling that can shape insecurity and shatter one's sense of confidence in the ground of one's existence. It is a situation which can make bodies react defensively as they reach out for support or search for a place to reground and re-orientate their relation to the world" (Simonsen, 2013, p. 20). On the other hand, performing places through disorientation can produce a more imaginative perception of where we are. Indeed, in Simonsen's view, this is the daily experience of migrants, engaged in a continuous meeting/clashing with new places and new cultures: "Moments of disorientation, then, can be seen as destabilizing and undermining, but they can also be seen as productive moments leading to new hopes and new directions. An example is migration, which can be described as a process of disorientation and reorientation as bodies both 'move away' as well as 'arrive'" (ibid.). Migration is an ancient phenomenon that has always been a part of human history. In the contemporary era, the migratory phenomenon has taken on a global dimension that touches the lives of most, whether as participants or spectators. Ultimately, "orientation/disorientation/reorientation in space and time, far from being a simple contextual backdrop *is* the story" (Baynham, 2003, p. 351).

In moving from one place to another, migrants are forced to engage with new cultures, new languages, and new body languages, with "multiple dislocations" and the "fragments of a broken geography", as in Meena Alexander's *Faultlines* (1993). Migrants may initially experience a total void, a feeling of total disorientation or of cultural shock: "This covers the period immediately after arrival and is a recurrent sense of loss. Disorientation arises from trying to come to terms with a new way of life, new customs, a new language, new administrative procedures, and even family role reversals" (Boyle et al., 2014, p. 192). Disorientation can cause a sense of maladjustment, of not belonging to either the native culture left behind or to the new culture; in turn this can give rise to fragmented personalities and a "consciousness of duality" (Anzaldua, 1987, p. 59) rooted in a space of ambiguity and contradiction. A new "mestiza consciousness" "shaped by cultural collision, forged out of mental and emotional states of perplexity and confusion" (Sandercock, 1998). Gloria Anzaldua narrates her own experience as a coloured woman, feminist, and lesbian in the space of the border zone between the United States and Mexico:

> A borderland is a vague and undetermined place created by the emotional residue of an unnatural boundary. It is in a constant state of transition. The prohibited and forbidden are its inhabitants. Los atravesados live here: the squint-eyed, the perverse, the queer, the troublesome, the mongrel, the mulato, the half-breed, the half dead; in short, those who cross over, pass over, or go through the confines of the "normal".
>
> (Anzaldua, 1987, p. 5)

For Anzaldua, disorientation involves having to continuously adapt to new cultural norms that reject her and to express herself in an unacknowledged hybrid language: "A new language – the language of the Borderlands. There, at the juncture of cultures, languages cross-pollinate and are revitalized; they die and are born. Presently this infant language, this bastard language Chicano-Spanish, is not approved by any society"[57] (ibid., Preface). In the ambiguous space of the *frontera*, the threat of violence is ever lying in wait for the migrant: "*La mojada, la mujer indocumentada* is doubly threatened in this country. Not only does she have to contend with sexual violence, but like all women, she is prey to a sense of physical helplessness. As a refugee, she leaves her familiar and safer home ground to venture into unknown and possibly dangerous terrain. *This is her home this thin edge of barbwire*" (Anzaldua, 1987, pp. 12–13). Furthermore, Anzaldua's sexual orientation places her amongst the ranks of the deviant, who are rejected by both the old and the new cultures: "The queer are the mirror reflecting the heterosexual tribe's fear: being different, being other and therefore lesser, therefore sub-human, in-human, non-human" (ibid., p. 18). Fear becomes an element of daily life in the Borderlands:

> The world is not a safe place to live in. We shiver in separate cells in enclosed cities, shoulders hunched, barely keeping the panic below the surface of the skin, daily drinking shock along with our morning coffee, fearing the torches being set to our buildings, the attacks in the streets. Shutting down. Woman does not feel safe when her own culture, and white culture, are critical of her; when the male of all races hunt her as a prey.
>
> (ibid., p. 21)

When migrants arrive at their destination, they need to find work, a home, and a new cultural frame of reference. In the process of reorienting themselves, they daily struggle with the challenge of reconstructing a divided personality, of figuring out who they are in a world that often no longer corresponds to their original social and physical coordinates: "Alienated from her mother culture, 'alien' in the dominant culture, the woman of color does not feel safe within the inner life of her self. Petrified, she can't respond, her face is caught between los intersticios [the interstices], the space between the different worlds she inhabits" (ibid., p. 20). Living in a place, which Heidegger referred to as *dwelling*, implies ongoing negotiation and renegotiation of our relationship with the space, places, people, and objects that surround us, in a continuous dynamic flow of encounter with the "familiar" and the "unfamiliar": "The question of orientation becomes, then, a question not only of how we 'find our way' but how we come to 'feel at home'" (Ahmed, 2006, p. 7). The migrant's experience of disorientation and reorientation has been taken up in contemporary literature, film, and TV series in a continuous decomposition and recomposition of places and languages. Zadie Smith narrates the stories of London immigrants from different countries of origin, who are divided between tradition, new expectations and novel social roles:

> Repetition […] that experience of moving from West to East or East to West or from island to island. Even when you arrive, you're still going back and forth; your children are going round and round. There's no proper term […] – original sin seems too harsh; maybe original trauma would be better.
>
> (Smith, 2000, p. 161)

For migrants who are continuously on the move, disorientation itself may become a familiar experience: "'getting lost' still takes us somewhere; and being lost can in its turn become a familiar feeling. Familiarity is shaped by the 'feel' of space or by how spaces 'impress' upon bodies" (Ahmed, 2006, p. 7). Thus, for Ahmed, places are almost a second skin that wraps itself around our bodies. Migrants are forced to continuously re-occupy this skin and readjust their bodies to new social conventions and, through the perceptions from their sense organs, to new landscapes, climatic conditions, tastes, sounds, and smells. In distant spaces, they search for roots and traditions to make them "feel at home": "If we are to feel that we belong, and have a place in the world, then we must be able to compose a suitable space. If we cannot 'situate' ourselves comfortably in relation to other people or things, we feel, quite literally, 'misfits'" (Davidson, 2000, p. 647). The dietary habits, traditional medicines, material cultures, and use of public space that characterize other cultural groups leave visible traces in the places to which their members migrate. Often unexpectedly, the landscape comes to feature "exotic" plants and herbs – such as the rows of chili peppers grown by Punjab Sikhs in farmsteads along the banks of the river Po in Northern Italy – crops, architectural elements, and sounds that are linked to the migrants' native cultures. These signs express loneliness for faraway places and objects which migrants think belong to them, but in reality only go on existing in their imaginations: "For the truth is that you can never simply 'go back', to home or to anywhere else. When you get 'there' the place will have moved on just as you yourself will have changed" (Massey, 2005, p. 124). In the new "home", almost always a city, encounters, narratives, memories, and exchanges continuously recreate the fabric of places and the people who live in them. Belonging becomes a problematic experience, destined to be renegotiated for generations. "Multicultural", "intercultural", and "hybrid" have become buzzwords, underpinning "assimilation" agendas that have little to do with the problematic nature of belonging and the highly diverse life stories of the people migrating to a particular area. The cover of a special issue of *Time* in 1993 featured a computer-generated image of the face of a woman and bore the title The New Face of America, or the woman of the future who exemplifies: "How Immigrants Are Shaping the World's First Multicultural Society". The myth of interculture presents the citizen of the future as a multiracial and multicultural hybrid who is nonetheless well adjusted to the consumer society: "Much of the language of enlightened cultural studies and the wider rhetorics of hybridity and unboundedness chime (sometimes all too easily) with the dominant tropes of neoliberalism" (Massey, 2005, p. 163). The "simultaneous heterogeneity" (ibid., p. 105) of people in movement brings us back to the concept of space: "If space is rather a simultaneity of stories-so-far, then places are collections of those stories, articulations within the wider power-geometries of space" (ibid., p. 130). "Space not as smooth surface but as the sphere of coexistence of a multiplicity of trajectories" (ibid., p. 63), an open discontinuous space in which we can trace multiple constructions of persons and objects, bound together by encounters, narratives, and representations. Doreen Massey elaborates: "It is in the happenstance juxtaposition, in the unforeseen tearing apart, in the internal irruption, in the impossibility of closure, in the finding of yourself next door to alterity, in precisely that possibility of being surprised [...] that the chance of space is to be found. The surprise of space" (ibid., p. 116).

Out of place[58]

We have seen that the experience of travel and arriving in a new place can give the migrant a recurrent sense of being lost. This sense of loss and the ambiguity implicit in the attempt to build oneself a new and fragmented belonging is a constant theme in the biographies of migrants, even after the first generation: "The disorientation of the sense of home, as the 'out of place' or 'out of line' effect of unsettling arrivals, involves what we could call a migrant orientation. This orientation might be described as the lived experience of facing at least two directions: towards a home that has been lost, and to a place that is not yet home" (Ahmed, 2006, pp. 19–20). In *Out of Place. A Memoir*, Edward Said recounted his own personal experience of disorientation as a psychological condition, starting from his very name: "Yet the overriding sensation I had was of always being out of place. [Thus it took me about fifty years to become accustomed to, or, more exactly, to feel less uncomfortable with 'Edward', a foolishly English name yoked forcibly to the unmistakably Arabic family name Said]" (1999, p. 3). Exile, travel, and displacement were all key elements in Said's life, which was perennially divided among different languages (Arabic, English, and French) and different places: "Along with language, it is geography – especially in the displaced form of departures, arrivals, farewells, exile, nostalgia, homesickness, belonging, and travel itself – that is the core of my memories of those early years. Each of the places I lived in – Jerusalem, Cairo, Lebanon, the United States – has a complicated, dense web of valences that was very much a part of growing up, gaining an identity, forming my consciousness of myself and of others" (ibid., p. XII). In this autobiographical essay, written with a sense of urgency, shortly before the author's death, Said described the feeling of being out of place as a key part of his personality, divided between "a record of an essentially lost or forgotten world" (ibid., p. IX) and his later intellectual and political experiences. This feeling of displacement was not only spatial but meant being out of place in his own skin, thus never feeling at home anywhere:

> I occasionally experience myself as a cluster of flowing currents. I prefer this to the idea of a solid self, the identity to which so many attach so much significance. These currents, like the themes of one's life, flow along during the waking hours, and at their best, they require no reconciling, no harmonizing. They are "off" and maybe out of place, but at least they are always in motion in time, in place, in the form of all kinds of strange combinations moving about, not necessarily forward, sometimes against each other, contrapuntally yet without one central theme [...] With so many dissonances in my life I have learned actually to prefer being not quite right and out of place.
>
> (ibid., p. 295)

The "reorientation" process may be complex and difficult to an extent that varies in line with the historical-political setting, subjects' social and economic skills or simply their bodies. Similarly, the ease with which migrants are "taken in" and "integrated" or "assimilated" also varies within the framework of a given logic of power:[59] "Space is fundamental in any form of communal life, space is fundamental in any exercise of power" (Foucault, 1984, p. 252). Thus diasporic space is laden with political meanings, while the process of reorientation may be interrupted, even after a number of years, when "unsuitable" citizens or bodies become newly visible, in response to political or racist logics, and are suddenly targeted as not belonging to the society hosting them: "Diasporic spaces do not simply begin to take shape with the arrival of migrant bodies; it is more that we only notice the arrival of those who appear 'out of place'" (Ahmed, 2006, pp. 19–20).

Abigail G.H. Manzella, in "Disorientation in Julie Otsuka's *When the Emperor Was Divine.* The Imprisoned Spaces of Japanese Americans during World War II", describes the US policy of marginalizing American citizens of Japanese origin and interning them in concentration camps during the Second World War. Manzella uses the expression *commutatory practices* to describe a targeted programme for isolating Japanese migrants implemented by the US government:

> In the specific instance of the 1940s U.S concentration camps, the government adopted a particular commutatory practice that I term disorientation because it resulted in the migrants' loss of a sense of place and self through a complex pattern of movement and alienation from space. They have lost their bearing. The government's policy to "disorient" by compelling the internees to turn away from the East – here it is not the compass direction but any Japanese cultural connections – and to

> instead adopt an imagined communal binary view of "us vs. them," opposing Americanness to Japaneseness. [...] When characters lose their sense of place they also lose their sense of self, resulting in self-hatred, disappearance, or vengeance.
>
> (Manzella, 2013, pp. 146–147)

Before leaving their US homes, during the long journeys from one place of internment to another, and in the camps themselves, the internees were obliged to give up every aspect of their Japaneseness, to change their lifestyles, language, and dietary habits, and to destroy all the objects connecting them to their country of origin: "None of those characters is merely relocated. Each is taken on a convoluted ride that promotes confusion through indirect paths and different means of transportation. This constant movement is part of the disorientation process. They can never be completely certain where they are going or when they will be moved next" (Manzella, 2013, p. 148). Giving up their Japanese culture was part of a process of forced assimilation: "Assimilation becomes a direct disorientation" (Manzella, 2013, p. 154). This process of marginalization and assimilation was irreversible. After their release, the Japanese migrants returned to find that their homes had been vandalized, they had lost their jobs, and they were instructed to keep a low profile to avoid being targeted for attack or victimization. "They are being trained to stay within their confines through both physical and psychological means; they are being taught to know their place" (Manzella, 2013, p. 151).

Feeling disoriented and displaced seems to be an increasingly common experience in contemporary society. In the words of Stiegler:

> This cardinal orientation is not successfully occurring today; thus we are suffering from disorientation as such. This leads above all to the speed of technical development since the Industrial Revolution, which has continued to accelerate, dramatically widening the distance between technical systems and social organizations as if, negotiation between them appearing to be impossible, their final divorce seems inevitable.
>
> (Stiegler, 2009, p. 3)

Feeling out of place is no longer the prerogative of migrants but is part of the malaise afflicting many who do not necessarily always identify with the imposed models of our "show-culture". Fashion, the media, and advertising on social networks impose rapid and continuous adjustment to stereotypical models of visibility and behaviour that do not reflect the everyday lives of many people: the contemporary myths and expectations of social and economic success, new professional competences in the areas of information technology and languages, a white body that is perennially young and healthy, and a free and prolonged sex life but at the same time a functional and happy family, all conceal narratives of inadequacy and solitude.

Notes

1 See Kant (1786).
2 Cfr. Stegmaier (2008, pp. 62–79).
3 Cfr. Kant (1998, Chapter III. "On the ground of the distinction of all objects", pp. 354–366 (in or. pp. 152–176)).
4 Cfr. Cassirer cit. in Farinelli (2003, p. 58).
5 Cfr. Stegmaier (2008, pp. 55–62).
6 Cfr. also Withers (2011, especially pp. 53*ff*).
7 Cfr. also "Both geography and anthropology were taught by Kant because of their 'pragmatic' dimension, the way this knowledge can guide us in our moral and practical life", Elden (2011, p. 3).
8 Cfr. *Nachricht von der Einrichtung seiner Vorlesungen in dem Winterhalbenjahre 1765–1766*, Akademie-Ausgabe Vol. 2, 1905, p. 310; cit. from, *Anthropologie*, 1983, p. 10, Anm.3.
9 Cfr. Immanuel Kant, *Von den Verschidenen Racen der Menschen*, AK 2, p. 433, cit. in Withers (2011, p. 52).
10 Kant (1992, "On space", §15, pp. 395–398).
11 Cfr. Withers (2011, 53*ff).*
12 Kant (1768), in Dünne and Günzel (2006, 75*ff).*
13 "A rule stands there like a sign-post. – Does the sign-post leave no doubt open about the way I have to go? Does it shew which direction I am to take when I have passed it; whether along the road or the footpath or cross-country? But where is it said which way I am to follow it; whether in the direction of its ringer or (e.g.) in the opposite one? – And if there were, not a single sign-post, but a chain of adjacent ones or of chalk marks on the ground – is there only *one* way of interpreting them? – So I can say, the sign-post does after all leave no room for doubt. Or rather: it sometimes leaves room for doubt and sometimes not. And now this is no longer a philosophical proposition, but an empirical one" (Wittgenstein, 1958, § 85, pp. 39–40).
14 "The term origo is introduced by Bühler (1934) and is conceptually understood as the origin of a coordinate system of subjective orientation (Bühler, 1934/1982a, p. 102), which is used to organize the personal, spatial, and temporal structure of utterances. Bühler (1990, p. 117) assumes one single origo for all dimensions, a mutual coordinate starting point for personal, local, and temporal deixis" (Ellen Fricke, "Origo, pointing, and speech – the phenomenon of two non-identical origos on the gestural and verbal level", a paper presented at the First Congress of the International Society for Gesture Studies, University of Texas at Austin, 2002, available on line at http://www.gesturestudies.com/files/isgsconferences/Contributions/Fricke/Fricke.html). Accessed 3 April 2017.
15 Cfr. Günzel, in Dünne and Günzel (2006, p. 29).
16 Cfr. also the in-depth discussion on the theme in van Cleve and Frederick (1991).
17 Cfr. Immanuel Kant: *Von dem Raume*, in Dünne and Günzel (2006, pp. 76–79); here p. 78 (emphasis in original). Kant (1992, "On space", §15, p. 397).
18 Ibid. (emphasis in original).
19 "Kant invokes transcendental schemata in order to explain how it is possible to subsume intuitions under pure concepts, how categories that are fixed a priori can be applied to changing a posteriori appearances" (Morris, 2004, p. 34).
20 See Dünne and Günzel (2006, p. 30).
21 Ibid.
22 Cfr. Dünne (2006, p. 30, emphasis in original) and Morris (2004, p. 113).
23 See Kant (1998).
24 Immanuel Kant, in Allison (1973).
25 Cfr. Dünne (2006, p. 30). Waxman (2014).
26 See James (1950).

27 Cfr. Morris (2004, p. 6).
28 Kant (1798, p. 51).
29 Dünne (2006, p. 30 passim).
30 Cfr. also Schroer (2006, pp. 40*ff*); see also 2.7.1.
31 The basis of lifeworld experience has been revisited in Herder's physiological-intrinsic concept of the space of perception, in Jakob von Uexkuells' biosemiotics – involving the subjective perception of space and time in the parallel cosmos of the tick –, and in von Bertalanffy's general systems theory.
32 Cfr. Dünne (2006, pp. 40*ff* and pp. 94–102).
33 Cfr. ibid. p. 41 and Löw (2001, pp. 69*ff*) and, specifically concerning gender relations in the space of social action, pp. 89–93; 173–179.
34 Numerous philosophers and writers, including for example Friedrich Schleiermacher and Goethe, took part in this debate.
35 Lessing had immortalized Mendelsson in his play *Nathan der Weise* (1779).
36 Cfr. Stegmaier (2008, pp. 66–77).
37 ibid., pp. 76*ff.*
38 Moses Mendelssohn: *Morgenstunden, oder Vorlesungen über das Daseyn Gottes* (1785), cit. in Werner Stegmaier (2005, pp. 9*ff*).
39 ibid., p. 10.
40 Original italics.
41 Cfr. Marcuzzi (2011, p. 130).
42 Cfr. Harvey (2007, p. 179*ff*).
43 Stegmaier (2005, p. 22).
44 Stegmaier (2008, cfr. p. 85); Cfr. also Morris (2004, p. 11): the "difference [between left- and right-handedness] [...] is tied to the body, rather than transcendental structures" and "in his metaphysical conception of the concept of space, he refers to the place [*Ort*] 'in which I find myself'"; similarly, in this same context of Kant's right-left distinction, Marcuzzi (2011, p. 129) understands "the body or bodies as a nonconceptual form".
45 These two examples reflect the reality of everyday life in a German city – namely, Königsberg, the main city in East Prussia, with a population of about 60,000 inhabitants in 1788 – in Kant's day, when both dwellings and streets were poorly lit.
46 Cfr. Immanuel Kant (1786, p. 80): "But in fact the faculty of making distinctions through the feeling of right and left comes naturally [...] – It is a faculty implanted by nature but made habitual through frequent practice" (Kant, 1996, p. 9).
47 Here, Kant was writing shortly after the publication of the first edition of *The Critique of Pure Reason* in which he had already begun to lay out his theory of knowledge. In *Metaphysische Anfangsgründe der Naturwissenschaft* (1786, written in parallel with his treatise on orientation) he transferred the Cartesian space of the outer world to the inner world of the transcendental subject. Cfr. Quadflieg (2009, p. 280).
48 Cfr. also Stegmaier (2008, p. 89).
49 This perspective is in keeping with Mozart's Enlightenment-inspired operatic work, *Die Zauberflöte*in [*The Magic Flute*], in which the Queen of the Night is defeated by reason, represented by the wisdom of Sarastro.
50 Cfr. Kant (1983, p. 47, italics in original).
51 This initial discovery of the DNA molecule by Miescher remained unknown, to the extent that DNA is generally associated with Francis Crick and James Watson who successfully identified its structure in 1953.
52 To dwell "is the basic character of Being in keeping with which mortals exist" (Heidegger, 1993, p. 362).
53 See Ahmed (2006, p. 27*ff*).

54 R. Braidotti, "Between the No Longer and the Not Yet: Nomadic Variations of the Body" (conference at the 4th Feminist European Conference, 29 September 2000) online at www.let.uu.nl/womens_studies/rosi/cyberfem.htm. Accessed 2 April 2017.

55 See Whatmore (2002) and Lejano et al. (2013).

56 See object-oriented ontology (OOO).

57 In this book, Anzaldua uses different varieties of English and Spanish, deliberately making it difficult to read for those who are not bilingual: this is to help readers appreciate her frustration with a monolingual world that rejects difference.

58 "Indeed, a sustained investigation of the 'out of place' metaphor points to the fact that social power and social resistance are always already spatial. When an expression such as 'out of place' is used it is impossible to clearly demarcate whether social or geographical place is denoted – place always means both" (Cresswell, 1996, p. 11).

59 "The mapping of particular meanings, practices, and identities on to place, they have argued, leads to the construction of normative places where it is possible to be either 'in place' or 'out of place'. Things, practices, and people labeled out of place are said to have transgressed often invisible boundaries that define what is appropriate and what is inappropriate" (Cresswell, 2009, p. 5).

Part II
Lost subjects

As we have seen, disorientation is a multidisciplinary theme that is informed by a wide range of sources. In the first section of the book, "Orientation/ Disorientation", I introduced disorientation as practice, situating it within the theoretical frames of geography. In the second and third chapters, I presented the theoretical/methodological framework of orientation/disorientation as a philosophical issue, from Kant to phenomenology and post-phenomenology, beginning with the philosophical question of what is space and finally entering into the post-phenomenological debate. The body and embodied experience are central to the theory and methods of phenomenology and post-phenomenology, which are informed by the thinking of Merleau-Ponty. This key role of the body is taken up again in feminist and post-colonial theory.

Thus, the second section of the book, "Lost Subjects", examines the theme of difference in the orientation of increasingly fragmented and multiple geographical subjects. At issue in post-structuralist theories in particular, differences are not given: rather disorientation may be "an effect of differences as well as a mechanism for their reproduction" (Ahmed, 2006, p. 60). Our perception of space is related to our "corporeal schema". Different subjects display different "ways of inhabiting and being inhabited by space" (ibid., p. 67). The theme of the geographical subject[1] brings us face to face with the key question of this book: how do we orient ourselves and why do we get lost? Why do some people have better spatial abilities than others? Do we have an internal compass and does it depend on culture, gender, race, age, physical or mental conditions? In the debate on orientation/disorientation, we are confronted by the ambiguity of purely neurobiological or cultural explanations of human behaviour and the complex functioning of our brains. I do not set out here to answer these questions or take a particular stance on them, but to outline issues and interpretations of disorientation in three different areas. In the three chapters of this section (Chapters 3–5), I consider disorientation in relation to gender issues (Chapter 3), the controversy about instinct in humans and non-humans (Chapter 4), and illness, dementia, drugs, and spatial disabilities (Chapter 5).

Note

1 See Butler (2015).

4 It doesn't matter which way you go

The lived body and disorientation[1]

> "Would you tell me, please, which way I ought to go from here?"
> "That depends a good deal on where you want to get to."
> "I don't much care where –"
> "Then it doesn't matter which way you go."
>
> (Lewis Carroll, *Alice in Wonderland*, 1865)

Gender relations and spatial tasks

In the course of the wide-ranging enquiry into disorientation, the subject of spatial experience has been called into question from a variety of perspectives. What does humans' ability or inability to orient themselves depend on? In Rosi Braidotti's terms, the human is itself a historic construction and normative convention that is

> instrumental to practices of exclusion and discrimination. […] The human norm stands for normality, normalcy and normativity. It functions by transposing a specific mode of being human into a generalized standard […] posited as categorically and qualitatively distinct from the sexualized, racialized, naturalized others and also in opposition to the technological artefact.
>
> (Braidotti, 2013, p. 26)

This single normative standard of the "human" lays down specific codes of behaviour and modes of being that are reflected both in our constructions of the body and in how we act, generally leading us to reject diversity and acknowledge only one model of subject:

> All other modes of embodiment are cast out of the subject position and they include anthropomorphic others: non-white, non-masculine, non-normal, non-young, non-healthy, disabled, malformed or enhanced people. This process is inherently anthropocentric, gendered and racialized in that it upholds aesthetic and moral ideals based on white, masculine, heterosexual European civilization.
>
> (ibid., p. 68)

In the course of the lengthy historical construction of this model, which is based amongst other components on a male/female binary logic, male superiority has also been claimed in the area of spatial dynamics. From the late nineteenth century onwards, numerous scholars (Darwin, 1873; Binet, 1894; van Gennep, 1909; Cornetz, 1909; Jaccard, 1932; etc.) set out to systematically explore orientation in humans and non-humans and to ask whether it was related to some kind of instinct (see Chapter 5). In the course of this enquiry, which is still ongoing today, interest arose in gender relations, understood in the dichotomous sense, and the factors and mechanisms implicated in them that might lead to differences in male and female ability to navigate space. Such differences, and the implicit superiority of men when it comes to orientation, may be used to justify practices of gender discrimination.

"Don't ask geographically challenged women to navigate. Reading maps and understanding where you are relies on spatial ability. Brain scans show that spatial ability is strong in males but poor in females. It's a male hunting skill" (Pease and Pease 1999, p. 63), write Allan and Barbara Pease in their famous and widely translated *Why Women Can't Read Maps … and Won't Stop Talking*. Pease and Pease's international bestseller presents a stereotyped and simplified account of the attempt to explain human behaviour in terms of innate biological characteristics: "The biological evidence now available […] shows convincingly that it is our hormones and brain wiring that are largely responsible for our attitudes, preferences and behaviour" (Pease and Pease, 2001, p. 6). In a paragraph entitled "Women don't have good spatial skills because they evolved chasing little else besides men", the authors back up this claim by stating that: "There are thousands of documented scientific studies that confirm male superiority in spatial skills" (ibid., p. 110). Although numerous field and laboratory studies (see Gilmartin and Patton, 1984; Montello et al., 1999, Zinser et al., 2004) have failed to confirm male advantage in the spatial or scientific domains, the Peases invoke a broad psychological and neurobiological literature as "scientific proof" of the popular belief that "women can't read maps", have greater difficulty in finding their way when driving, and get lost more easily than men. A myriad of experimental and field studies – many of which have been interdisciplinary and comparative – conducted over half a century in the domains of psychology and neurobiology, would appear to justify the conclusion that while women display superior verbal ability, men behave more aggressively and are more skilled in mathematical and visuospatial activities. To cite but a few examples: "Virtually all genuinely spatial tasks show a significant and substantial male advantage" (McBurney et al., 1997, p. 165); "The existence of gender differences in favour of males in spatial abilities has been clearly established" (Voyer et al., 2000, p. 1). Doreen Kimura adds:

> Scientific evidence for consistent differences in cognitive function between men and women has accumulated for well over 50 years. […] A solid body of research, carried out primarily in North America and Western

> Europe, has established that men, on average, excel on spatial tasks […], perception of the vertical and horizontal, mathematical reasoning, and spatio/motor targeting ability. Women, on average, excel on tasks of verbal fluency […], perceptual speed, verbal and item memory, and some fine motor skills.
>
> (Kimura, 1998, p. 155)

This is but a small sample from a vast field of enquiry that has investigated spatial visualization and orientation in men and women (Maccoby and Jacklin, 1974). Differences have been variously attributed to brain lateralization, genetic makeup, male sex hormones – which are positively correlated with spatial abilities – and the size of the hippocampus. According to the "two-brain" theory, as Pease and Pease's simplified version would have it, the left hemisphere is specialized in verbal, analytical, and sequential processes, while the right hemisphere tends to be dominated by spatial, holistic, and simultaneous cognitive functions: female behaviour is claimed to be more influenced by the left or verbal hemisphere, and male behaviour by the right or visuospatial one. On top of these factors come differential socialization experiences and identification with gender roles (Voyer, 2000). Gender differences in the spatial sphere also appear to affect wayfinding strategies, with women preferring to use landmarks as a guide and men favouring absolute reference points such as the cardinal directions:

> Men were found to be more likely than women to report using an orientation strategy of way finding, wherein one's own position is tracked in relation to geographical reference points. Women were more likely than men to report using route strategy, with the focus on learning the features of a specific route, particularly those features where a change in direction is required.
>
> (Lawton, 1994, p. 772)

The evidence for these conclusions includes the outcomes of a carefully thought-out and cautiously interpreted study by Bever (1992), who set out to investigate the genetic bases of neurophysiological differences in the spatial domain. However, equal care and caution is not always applied when the huge body of scientific data currently available is used to confirm the common stereotype of superior spatial skills in males with respect to females, in a prime example of "scientific racism". As Bever himself warns:

> The study of cognitive differences between genetically coded groups in the normal population is just one method to reveal cognitive distinctions that correspond to genetically coded neurophysiological differences. This method must be used with great social and political responsibility: The goal is not to "scientifically" define cognitive differences in general abilities […]. As our theories of the mind and the brain improve, it is not

> impossible that we will find an increasing number of cases of difference in cognitive style associated with biologically defined differences in populations. We must squarely face the possibility that such discoveries, like all of science, can be misused for political ends [...] to justify a form of scientific racism.
>
> (Bever, 1992, p. 207)

The self-help manual by Pease and Pease – an expression of 1990s pop psychology and the renewed *backlash* against women and feminism (Staszak, 2005) – bases its starting assumptions on an analysis "of more than a million boys and girls to study their spatial ability" (Pease and Pease, 2001, p. 109), attributed to Camilla Benbow, educational psychology professor and dean at Vanderbilt University, as well as a prominent member of the *National Mathematics Advisory Panel* set up by George Bush in 2006. Camilla Benbow is well known for her 1980s publications reporting some 80 intrinsic gender differences that put males at an advantage in the domain of advanced mathematical learning. However, Benbow herself has revisited the theme of difference between males and females in the field of scientific learning, reaching far more cautious and nuanced conclusions in a consensus statement written with other scholars: "There is no single factor by itself that has been shown to determine sex differences in science and math. Early experience, biological constraints, educational policy, and cultural context each have effects, and these effects add and interact in complex and sometimes unpredictable ways" (Halpern et al., 2007, p. 41). Based on a thorough review of the state of the art in research on sex differences, the authors of the statement point out – at the risk of failing to meet media and public expectations of sensationalist findings – that such a complex area defies the quest for straightforward explanations:

> There cannot be any single or simple answer to the many complex questions about sex differences in math and science. Readers expecting a single conclusion – such as that we can explain sex differences in science and math by knowing about hormones, or by knowing how stereotypes affect performance, or by knowing how our ancestors met the challenges in their lives – are surely disappointed.
>
> (ibid.)

The attempt to come up with a "scientific" explanation of "difference" based on biological/genetic factors is deeply rooted in the "grid of binary oppositions" (man/woman, body/soul, emotion/reason) that has dominated Western rationalism: "We find again here this disease of thinking in essences, which is at the bottom of every bourgeois mythology of man (which is why we come across it so often)" (Barthes, 1957, p. 114). This dualism is reflected in the domain of scientific and geographic learning, in which academically coded knowledge is presented as the only possible form of "truth": "Geographical knowledge proclaims itself to be singularly true, but in so doing depends upon its difference from other 'false' knowledges. And, as a result of the meeting between this critique and feminism, it becomes clear that male dominance in the production of geographical knowledge is closely linked to the phallocentrism of its discourse" (Bondi and Domosh, 1992, p. 204). The work of Bondi and Domosh is informed by post-structuralist and deconstructionist perspectives that bring into question the very foundations of geographical knowledge. Cultural studies and feminist analysis challenge the concepts of authority and representation, rejecting the conventions and Western cultural hegemony of the modern era and interpreting difference as a complex and contradictory phenomenon that is unstable and influenced by logics of power.

The binary opposition between "man" and "woman" is conceptualized as a mechanism that, in practice, assists in constructing and legitimating gender difference: "It is neither inherent nor inevitable and serves to conceal experiential differences *among* men and women" (ibid., p. 201). The woman/man subject, in Braidotti's view: "is not a monolithic essence that has been defined once and for all, but rather the place of multiple, complex and potentially contradictory experiences, a place defined by overlapping variables such as social class, race, age, lifestyle, sexual preferences and so on" (Braidotti, 2002, p. 13). The logic of dualism, in contrast, imposes a normative and hierarchical framework, producing a "normality" that is the same for all and delimited by clear boundaries (race, age, sex). What all diversity has in common is the distance of bodies from this normality, bodies that "represent a play of reflections – and often, possibly distorting, mirrors – that stack up to form the parameters of what is acceptable. The norms emerging from this process satisfy the expectations of the phallocentric regime, which stakes its all on a body that is docile, reproductive, white, heterosexual and normally formed" (Braidotti, 1996, p. 11).

The feminist critique's unpacking of the subject has led to a questioning of "woman" and "man" as universal categories, given that being male or female is woven into a web of other types of belonging, such as class, ethnic group, sexual orientation, etc. In this context, even the concept of gender becomes ambiguous. As Judith Butler puts it: "Perhaps this construct called 'sex' is as culturally constructed as gender; indeed, perhaps it was always already gender, with the consequence that the distinction between sex and gender turns out to be no distinction at all" (Butler, 1999, p. 11), Young, drawing on a phenomenological-existentialist approach inspired by Simone de Beauvoir and Merleau-Ponty, supports Moi's proposal to substitute the concept of gender with that of *lived body* and to conceptualize the body as a *situation*:

> The idea of the lived body thus does the work the category "gender" has done, but better and more. [...] the category of the lived body allows description of the habits and interactions of men with women, women with women, and men with men in ways that can attend to the plural

> possibilities of comportment, without necessary reduction to the normative heterosexual binary of "masculine" and "feminine".
>
> (Young, 2005, p. 18)

The notion of the lived body also overcomes the distinction between nature and culture that is intrinsic to the sex/gender dichotomy and justifies hegemonic male discourse. The lived body: "is a unified idea of a physical body acting and experiencing in a specific sociocultural context; it is body-in-situation" (ibid., p. 16). From birth, our bodies are shaped by conventions and stimuli that are rooted in the different cultures and reflected in our mode of speech, dress, familial role, occupation, hobbies, and propensity to engage in given physical or intellectual activities. Orientation is a key component of the lived body theorized by Young: "Even in the most simple body orientations of men and women as they sit, stand, and walk, we can observe a typical difference in body style and extension" (ibid., p. 32).

Hunters and gatherers?

Within the framework of gender relations, do differences, of whatever kind, depend on "nature" or "culture"? The nature–nurture debate in relation to spatial abilities, is the starting point of all discussions on sexual belonging and orientation in the human species. This is the realm of the famous "hunter–gatherer" theory, which blends, often in a confused manner, explanations based on the primitive origins of humankind and the stereotype of natural abilities linked to sexual difference: "Within ancestral environments females mainly specialized in gathering activities and males in hunting" (Ecuyer-Dab and Robert, 2004, p. 11). Laboratory studies on rodent and primate species suggest a model of sexual selection based on wider ranging by males in search of females and food (Gaulin and Fitzgerald, 1989) and on the larger size of the male hippocampus, the brain part associated with spatial processing (see Chapter 2). During hunting, human males are thought to have drawn on Euclidean and geometric coordinates to orient themselves (Galea and Kimura, 1993), and to have developed superior ability to strike a target (Thomas and French, 1985). Anatomical characteristics, such as greater muscle mass (the "broad shoulders" of popular culture), aggressive behaviours and competition among males, provided reproductive advantage thereby contributing to the improvement of the species. This prompts the interesting claim that "Men ordinarily engage in more active courtship than women" (Ecuyer-Dab and Robert, 2004, p. 22).

In women, on the other hand, a more limited range of movement is thought to have favoured a greater propensity to attend to the immediate environment, which in turn would explain the development of orientation strategies based on landmarks (Ecuyer-Dab and Robert, 2004). Thus, men and women are proposed to have adapted to different roles within the framework of an evolutionary reproductive model. Laboratory experiments also suggest that, in the course of evolution, our male and female ancestors may have developed specialized roles in the sourcing of food supplies, with the division of labour dating back to the Plio-Pleistocene period (Eals and Silverman, 1994). Our female forebears are thought to have mainly looked after the procurement of plants, fruit, eggs, insects, and small mammals, with a low level of associated risk. Such a system might also have been underpinned by physiological factors, with the ties of female fertility and maternity impacting negatively on women's spatial behaviour and knowledge of orientation, but promoting their reproductive success.

"How do girls and women constitute their experienced world through their movement and orientation in places?" (Young, 2005, p. 9) enquires Young in *Throwing Like a Girl*, while revisiting Ewin W. Straus' phenomenological study, *The Upright Posture,* on the significance of lateral space in the erect position. Straus had observed a "remarkable difference in the manner of throwing of the two sexes" (Straus, 1966, p. 157), attributing this to a "feminine attitude" towards throwing. According to Straus the difference is biologically based, but not specifically anatomical: "Girls throw in a way different from boys because girls are 'feminine'" (Young, 2005, p. 28). Furthermore, this "feminine attitude" is encouraged from the first year of life by the way children are reared: there is a relationship between the practice of sport and spatial performance in boys and girls, while toys classified as "male" and "female" typically offer different levels of spatial content (cars and tricycles versus dolls and toy kitchens) (Gilmartin and Patton, 1984). In Young's view, in assuming her status as feminine, a girl "takes herself to be fragile and immobile and [...] she actively enacts her own body inhibition" (Young, 2005, p. 44).

In the practice of sport as in every other activity, the "feminine attitude" stems from the ambiguous intermingling of nature and culture: "There is a specific positive style of feminine body comportment and movement, which is learned as the girl comes to understand that she is a girl" (ibid., p. 43). Girls, more so than boys, have traditionally been invited to take care with their appearance, look after their clothes, avoid getting dirty, refrain from engaging in violent play, stay in the environs of home, etc. Not only does there exist a typical style of throwing like a girl: "but there is a more or less typical style of running like a girl, climbing like a girl, swinging like a girl, hitting like a girl" (Young, 2005, p. 33). Drawing on Merleau-Ponty's notion of motility, Young argues that female disadvantage in throwing and the spatial domain is due to what she terms an *inhibited intentionality* that prevents full mastery over space and movement: "To the extent that feminine bodily existence is an inhibited intentionality, however, the same set of possibilities that appears to be correlative to its intentions also appears to be a system of frustrations correlative to its hesitancies" (ibid., p. 38).

The hunters–gatherers theory leads us to the heart of the debate on how human societies may be influenced by environmental factors, a theme that is not new to geography and which remains of vital interest and worthy of serious ongoing reflection. Voyer and colleagues have argued that, even if we accept that the environment affects society, we should take into account that recent changes in male and female roles may have influenced gender differences in spatial abilities. Nonetheless, these same authors point out that there is little published research on the topic, suggesting that this apparent lack of interest may be linked to difficulty in obtaining funding and in publishing work on the "null hypothesis", that is to say, on *not* having found significant gender differences in the performance of spatial activities (Voyer et al., 2000). Pease and Pease do not appear to have been afflicted by any such doubts, providing a peremptory account of women's presumed inferiority in the spatial sphere: "Most women don't enjoy spatial activities and don't pursue careers or pastimes that require them. [...] Spatial ability is not strong in women and girls because being able to chase animals and find the way home was never part of woman's job description. This is why many women have trouble reading a map or street directory" (Pease and Pease, 2001, p. 110). As much as to say that women should be kept "barefoot and pregnant".

The propensity to get lost is not only viewed as a negative trait that reflects, yet again, the presumed "inferiority of women", but also bears moral valence: the "lost woman" is cast out from the society of "respectable" women on account of her immoral conduct; she becomes a "woman of the street", thus acquiring, through her "perdition", a new relationship with external space and orientation. The street can also assume the guise of a heterosexual space in which "sexual dissidents" are not authorized to show themselves: "Repetitive performances of hegemonic asymmetrical gender identities and heterosexual desires congeal over time to produce the appearance that the street is normally a heterosexual space" (Valentine, 1996, p. 150). The heterosexual street becomes threatening to difference of all kinds and fear continues to act as a severe constraint to movement: "The humiliation comes in acknowledging that my butch drag is not black enough, not leather enough, to hide my fear" (Munt, 1995, p. 115), comments Sally Munt in relation to her own experience as a lesbian on the streets of Nottingham.

Anxiety and fear contribute to shaping differences in orientation strategies and levels of apprehension about getting lost. Women are reported to display less independence in orienting themselves: "Partly because they perceive themselves at greater risk, of being raped, harassed, or otherwise victimized [...] in more distant locations" (Zinser et al., 2004, p. 664). This leads us once more into the geography of emotions and fear (Valentine, 1989; Pain, 1991; Davidson, 2003; Davidson et al., 2005; Smith, 2009) and the *emotional turn* in the social sciences (Bondi, 2005), as exemplified by Joyce Davidson's phenomenological analysis of agoraphobia (2003). For a woman, the threat of sexual violence represents an "invasion of her body space" that can take the form of being looked at or touched by a man: "Women tend to project an existential barrier closed around them and discontinuous with the 'over there' in order to keep the other at a distance" (Young, 2005, p. 43).

The fear of violence has also been argued to underpin protective behaviours that entail restricting the spatial experience of female children compared to that of males: boys are given more encouragement to move and are allowed a greater range of movement than that conceded, for safety reasons, to girls. Girls are taught to limit their movements, take care not to get dirty, keep their clothes in order, and avoid getting hurt, thereby learning that most things they would like to do are dangerous for them. A girl: "Thus, […] develops a bodily timidity that increases with age. In assuming herself to be a girl, she takes herself to be fragile" (ibid.). The threat of sexual violence as an explanation for women's alleged lesser inclination to move away from the home can help to justify the segregation and exclusion of women in sexist and patriarchal societies. Ardener argued that while the "raped woman" remains a popular theme in literature and film, in real life "rape is still exceptional" (Ardener, 1981, p. 23), and we might also ask why, historically, women have never been encouraged to learn self-defence techniques instead of depending on men for protection. Intercultural comparison suggests that a range of different factors may have conditioned differences in mobility between men and women:

> Mobility is not only, to some degree, determined by the nature of physical space; it also affects the appreciation of space. […] There have been a number of social practices which have probably contributed, whether advisedly or indirectly, to the greater restriction of movement in space of some (possibly minority) groups of women. Foot-binding, tight corseting, hobble skirts, high heels, all effectively impede women's freedom of movement, and make them dependent on mechanical or other forms of transport. […]. For instance, if women are not encouraged to ride bicycles or horses, to paddle canoes or to learn to drive cars, or to own these means of transport, their freedom to enter spaces may be relatively curtailed in comparison to men's. […] Another inhibitor may be an ideology which encourages women to be physically frail, or to think that they are.
>
> (Ardener, 1981, pp. 21–22)

Another factor that has been cited as evidence for the role of "nature" in influencing human behaviour is the theory of "preparedness" in the face of threat, defined as a predisposition to learn in response to the stimulus of fear. In evolutionary terms, greater anxiety about disorientation, getting lost and open spaces (agoraphobia) (see Chapter 6) might be explained by a female tendency to favour strategies for protecting offspring, making women less inclined to run risks: the survival of the children depends on that of the mother (Ecuyer-Dab and Robert, 2004). In other words, the role of maternal care, gestation, and breastfeeding in infant survival may have led women to place a high value on protecting their own lives (Campbell, 1999).

Nature and nurture may combine to influence women's freedom of movement and experience of the world: "A relationship between travel experience and gender could indicate that geographic knowledge is a product of nurture. As noted previously, girls and women may travel less and over shorter distances because of general anxiety about their security" (Zinser et al., 2004,

p. 668). Hormones have also been suggested to play a key role in "curbing female travelling" (Ecuyer-Dab and Robert, 2004, p. 18). Ecuyer-Dab and Roberts based this statement on the observation of animal behaviour (rats and monkeys) and on the analysis of cases observed across a range of tribal societies in Africa, Australia, and the Amazon rainforests, concluding – via a jump in logic which is audacious to say the least – that: "Even in a modern city of North America, we found that the home range travelled through daily for professional and/or personal purposes was larger in men than in women" (ibid., p. 23).

Again, Pease and Pease offer an insight into the popular beliefs feeding into the construction of gender stereotypes: ("Our basic biology and how our brains are organised are the culprits", Pease and Pease, 2001, p. 132). Indeed, we cannot ignore the possibility that these stereotypes may themselves influence orientation and spatial ability: "If the culture communicates that women are inferior to men in geographical knowledge, that can lower confidence in women or produce what has been called stereotype threat [...] which can contribute to inferior performance in women on geographic tasks" (Zinser et al., 2004, p. 682). The "threat of the stereotype", for which both laboratory studies and real-life case studies have provided evidence, undermines confidence in the members of a group viewed as lower performing (based on racial and gender stereotyping), making test candidates feel insecure in relation to their own abilities and therefore obtain lower scores (Spencer et al., 1999). According to Cornell, Sorensen, and Mio, despite the absence of significant gender differences in orientation ability, field experiments suggest that women rate their own sense of direction more poorly than do men, leading them to be less successful on orientation tasks (Cornell et al., 2003).

Finally, it is worth reflecting on whether and how the behaviour of animals (from primates to rodents, as we have seen in the case of orientation) and "proto-hominids" may be taken as a basis for explaining – and legitimating – other contemporary models of social and political organization (male dominance, supremacy of the leader, the maternal instinct, subjugation of the weak). As Staszak has argued: "Those who try, on the basis of respect for 'natural' differences, to justify inequality between genders are also those who, on the basis of different predispositions in individuals, try to legitimize a society governed by the law of the strongest" (Staszak, 2005, p. 15).

Women can't read maps?

If "Research suggests that men perform better than women on various spatial perception tasks" (Zinser et al., 2004, p. 661), how might such a difference come into play in geographical research? In face of the "popular wisdom" propounded by Pease and Pease, Hancock provocatively queries: "Are those women who insist on doing geography going against nature by venturing into a domain that they are not biologically equipped to tackle? On the contrary, it seems to me that it is rather the way in which the methods and concepts of geography are constructed that makes it a 'masculine territory' in which women struggle to create a 'place' for themselves" (Hancock, 2004, p. 167). The dynamics of the lived body in perceptions of space translate into a logic of exclusion as Luce Irigaray has explained: "It's not that we have a territory of our own but their fatherland, family, home discourse, imprison us in enclosed spaces where we cannot keep on moving, living as ourselves" (Irigaray, 1985, p. 212). More generally, spatial exclusion was part of the historical process that led to the creation of universities and modern science: "The establishment of those places was bound up with the distinction of gender and the exclusion of women" (Massey, 2005, p. 144).

In light of a large body of research, Zinser, Palmer, and Miller concluded that the influence of hereditary factors on geographical learning and knowledge of geographical sites is minimal: "Men and women learn, or can learn, these sites equally well and with equal ease" (Zinser et al., 2004, p. 681). Gilmartin and Patton also observed that differences in spatial abilities between males and females were limited in all cases (average standard deviation of 0.4) and inexistent in relation to the use of maps, in contrast with over-simplistic "popular axioms" claiming male advantage: "The generalization that males' spatial abilities are superior to females' does not hold in this context" (Gilmartin and Patton, 1984, p. 616). The authors thus conclude that: "There will be a need to eliminate some inaccurate stereotypes regarding 'natural' abilities of men and women" (ibid., p. 605). Montello and colleagues, on their part, based on a review of over 30 years' neuropsychiatric and psychometric research that had sought to explain male–female differences in spatial ability in terms of biological theories and/or theories of learning and socialization, concluded that: "Statistically reliable sex-related differences favoring males in their performance of several psychometric tests of spatial ability have repeatedly been found, though the differences are typically modest in size" (Montello et al., 1999, p. 516). However, they refrained from generalizing about gender differences in cartographic skills, arguing that we first require a clear definition of what we mean by "map".

Feminism and postmodernism both advance a deconstructionist critique of the map and of atlases, viewing them as instruments of power and of the transmission of an "objective science" that overlooks social and political implications (see Chapter 1). Thus, cartographic representations do not correspond to "reality", but to an arbitrary viewpoint; Harley argued that "Cartography embodies [...] a systematic social inequality. The distinctions of class and power are engineered, reified and legitimated by means of cartographic signs. The rule seems to be 'the more powerful, the more prominent'" (Harley, 1989, p. 245). Thus, the map does not tell us where to go but presents us with a given vision of the world: for Harley, maps are "authoritarian images" in that they all "employ the common devices of rhetoric such as invocations of authority" (ibid., p. 252). Maps – in the words of Mitchell – are: "Enigmas, problems to be explained, prison-houses which lock the understanding away from the world. [...] the sort of sign that presents a deceptive appearance of naturalness and transparency concealing an opaque, distorting, arbitrary mechanism of representation" (Mitchell, 2013, p. 8). However, the ability to read a map is implicitly underpinned by acceptance of a normative principle. Hence, the deliberately provocative question thrown out by Allen: "Because they are artifacts, maps must reflect certain aspects of the modes of thought underlying their creation. Is it possible that map reading as an activity taps the vestiges of male-characteristic cognitive tendencies expressed in the historic conventions associated with map design and use?" (Allen, 2000, p. 17). It follows that disorientation may also be understood as a method of resistance.

The relationship between women and space, which is at the heart of Shirley Ardener's thinking, demonstrates that people and objects define and influence the space in which they are situated and are in their turn influenced by it (like the pieces on a chessboard). Space, like gender, is an ordering category; no map (Ardener gives the London Underground as an example) actually corresponds to what we see:

> When dimension or location are introduced we assert a correspondence between the so-called "real" physical world and its "social reality". [...] Measurements, and what is measured, for instance, are neither totally imperative nor just random; choice enters "reality". Societies have generated their own rules, culturally determined, for making boundaries, and have divided the social into spheres, levels and territories.
>
> (Ardener, 1981, pp. 1–2)

Ariadne, lady of the labyrinth

Disorientation can express a psychological frame of mind, a state of distress, as narrated by Simone De Beauvoir in relation to her own personal experience:

> Cut off from everybody, I no longer had any link with the world: it was becoming a spectacle that did not concern me personally. At times, I lost all sense of reality: streets, cars, passers-by – all were but a procession of appearances amongst whom my nameless presence floated. [...] it is a very short step between utter loneliness and madness. I had good reason to lose my wits. For two years, I had been struggling to get out of a trap but without finding a way; I kept bumping into invisible obstacles: and this would give me a sense of vertigo.
>
> (De Beauvoir, 1958, p. 340)

In such cases, disorientation becomes a mental state, whether temporary or permanent, that reflects a sense of being out of place, of not being able to cope with the world in which one lives, a condition that is often experienced by the migrant (see Chapter 3): "A bodily feeling of losing one's place, and an effect of the loss of a place: it can be a violent feeling, and a feeling that is affected by violence, or shaped by violence directed toward the body" (Ahmed, 2006, p. 160). In the institutionalized model of official cartography, orientation is implicitly underpinned by a rational visualization of space, based on carefully chosen objectives and destinations, and precise timetables, directions, and boundaries: in contrast, if we are to get lost, we require the skill of "thinking without a banister", as advocated by Hannah Arendt, and the will to abandon conventional frameworks. Disorientation might then be interpreted as a different and freer relationship with space, a space that is "extraordinarily convoluted, multiply overlaid, paradoxically pleated, folded, broken and, sometimes, absent" (Rose, 1999, p. 247), and that may be perceived with the entire body and the imagination.

The "scientific precision" of many studies and laboratory experiments might thus be exchanged for the powerful and eternal metaphor of the labyrinth. A "rhizomatic" labyrinth which is entirely open, invariably accessible, acentric, and non-hierarchical (Deleuze and Guattari, 1987) (see Chapter 7), and can become a cartographic strategy for leading us through the complexity of contemporary existence, helping us to find our way around it, familiarize with its spaces, overcome our fear of it and devise methods of addressing our sense of spatial disorientation: "So that we may position ourselves as feminist intellectuals, as travellers in hostile countries, armed with maps that we ourselves have made, ready to follow paths that are often only visible to us, but which in compensation we can narrate, share and give an account of" (Braidotti, 2002, p. 124). The labyrinth from which only Ariadne, using a traditionally "female" resource, a ball of thread, successfully finds her way out. This makes her a "nomadic subject", as conceived by Rosy Braidotti: "The nomadic state is less defined by the act of travelling than by an awareness that feeds into the desire to overturn established conventions: it is a political passion for transformation or change" (ibid. p. 14). In her guise as a feminist intellectual, Ariadne faces no mean task, as Linda McDowell has observed: "The construction of partial and situated knowledge from a critical position will not be an easy task" (McDowell, 1992, p. 413). Such a task involves conceiving a new cartography and knowing how to interpret its signs, based on: "A way of thinking that completely dissolves the notion of a centre and consequently all notions of place of origin, true identity or the right way to get somewhere" (Braidotti, 2002, p. 16). Thus, "nomadic cartographies" are "constantly redrawn and as such by their very nature do not tolerate fixity nor greedy appropriation. The nomad has a strong sense of territory that does not translate into possessiveness" (ibid., p. 57).

The metaphor of *Getting Lost* is again invoked in the title of an essay by philosopher and feminist intellectual Patti Lather, which deconstructs the practice and methods of feminist research with a view to "coming to terms with the complexities involved in the 'post' turn in the human sciences" (Lather, 2007, p. 5). This "book about a book" was written during and after the writing of Lather's earlier *Troubling the Angels* (co-authored with Chris Smithies, 1997), an experimental ethnographic work on the lives of women in Ohio with HIV/AIDS, in which the author had set out to shed light on the difficulties, errors, contradictions, false starts, and blind alleys of field research: "I introduce what getting lost might mean as both a methodology and a mode of representation. At its simplest, getting lost is something other to commanding, controlling, mastery. [...] Lack, lost, missing, absence" (ibid. p.11). And here Lather attempts to address the issue of scientific racism flagged by Bever. Being lost:

> includes the loss of an innocent science too. Other losses include: the unitary, potentially fully conscious subject; researcher self-reflexivity as a "way out" of impasses in ethics and responsibility; and transparent theories of language. Lost also is the clear political object. [...] the concept of getting lost functions as a paradox. It is a means of critiquing a certain confidence that research must muster in the audit culture. It is a metaphor for a new generation of postcritical work.
>
> (ibid., p. 12)

In *Plateau 5: Loss and Lost, Lost, Lost*, Lather opens by citing Derrida: "In a certain way, there is perhaps no voyage worthy of the name except one that takes place there where, in all senses of the word, one loses oneself, one runs such a risk, without even taking or assuming this risk: not even of losing oneself but of getting lost" (Derrida in Lather, ibid.), offering her own view that:

> a stance of "getting lost" might both produce different knowledge and produce knowledge differently [...] Here, accepting loss becomes the very force of learning, and what one loves when lovely knowledge is lost is the promise of thinking and doing otherwise. [...] in negotiating what it means to (re)think critique and practice in such "dark enough" times [...] This book explores what it might mean to claim getting lost as a methodology for such times.
>
> (ibid., p. 13)

In sum, in Lather's discourse, getting lost remains a powerful metaphor, but also becomes a research strategy for engaging with the contradictions of feminist enquiry.

Sexual (dis)orientation

> And when one of them meets with his other half, the actual half of himself, whatever his orientation, whether it's to young men or not, the pair are lost in an amazement of love and friendship and intimacy, and one will not be out of the other's sight, as I may say, even for a moment.
>
> (Plato, *Symposium*)

Much scholarship has explored the origins of difference, which appear in great part to lie in neo-positivist and pseudoscientific accounts based on a mythical original model, identical for all, and on conformity to well-established roles. In contrast, Braidotti has argued that the body, or the embodiment of the subject: "is to be understood as neither a biological nor a sociological category but rather as a point of overlap between the physical, the symbolic and the material conditions" (Braidotti, 2003, p.44). In this construction of the body as geographical subject, a phenomenological approach can shed light on the relationship between space and sexuality. The dividing of space based on a heteronormative logic, which is particularly evident in the places of power, entails the exclusion of "sexual others" who consequently find themselves to be "out of place" (Hemmings, 2002; Hubbard, 2008).

"Normality" is hence normative and hierarchical, the same for all, and fixed by clearly defined limits. Diversity manifests itself in the distance of bodies from this normality. Sexuality is spatially configured not only in the sense that it inhabits space, but also in that bodies are constructed in terms of their particular mode of inhabiting space: "The body orientates itself in space, for instance, by differentiating between 'left' and 'right', 'up' and 'down', and 'near' and 'far', *and this orientation is crucial to the sexualisation of bodies.* Phenomenology helps us to consider how sexuality involves ways of inhabiting and being inhabited by space" (Ahmed, 2006, p. 67, original italics).

Sexual orientation contributes to the formation of the subject and to the spatial dimension of the identity construction process. The notions of orientation and disorientation are at the heart of Ahmed's phenomenological discourse. Her interest in dis/orientation stems from her concern with the issue of sexual orientation and the orientation of desire, leading her to apply phenomenology to queer studies: "What does it mean for sexuality to be lived as orientated? If orientation is a matter of how we reside in space, then sexual orientation might also be a matter of residence; of how we inhabit spaces as well as 'who' or 'what' we inhabit spaces with" (ibid., p. 1). As we have seen, emotions and intentionality are key to phenomenological and post-phenomenological discourse, and emotions are directed *towards* objects: "So when we feel fear, we feel fear *of* something. I brought this model of affect as contact: we are affected by 'what' we come into contact with. In other words, emotions are directed to what we come into contact with: they move us 'towards' or 'away' from such objects" (ibid., p. 2).

The term *sexual orientation* "coincides with the production of the 'homosexual' as a type of person who 'deviates' from what is neutral" (ibid., p. 69), from the "naturalness" of heteronormativity. To have a given sexual orientation means to be "directed" in one direction rather than another. The concept of "sexual orientation" is then extended to the themes of transgression and spatial transformation, bringing to light, through the experience of transgender and intersexual individuals, the problematic nature of all categorization and the fact that the geographical subject: "does not neatly divide into male/

female (or black and white, young/old, able-bodied/disabled etc.)" (Hubbard, 2008, p. 4). Thus, Ahmed has offered a reflection on the etymology of the word "queer" which derives from the Indo-European term "twist": "Queer is, after all, a spatial term, which then gets translated into a sexual term, a term for a twisted sexuality that does not follow a 'straight line', a sexuality that is bent and crooked" (Ahmed, 2006, p. 67); in other words, a sexuality that is disorientated.

Ahmed takes her own personal experience as her point of departure: "One of the reasons that I became interested in the very question of 'direction' was because in the 'middle' of my life I experienced a dramatic redirection: I left a certain kind of life and embraced a new one" (ibid., p. 19). Her sense of disorientation derives from her abandonment of the path marked out by the heterosexual universe: "Becoming a lesbian taught me about the very point of how life gets directed and how that 'point' is often hidden from view. Becoming reorientated, ... involves the disorientation of encountering the world differently" (ibid., p. 20). Sexual reorientation affects the entire positioning of the world, the sphere of what we feel to be familiar or unfamiliar: "Knowing which way we are facing, is about the making of worlds." Thus, getting lost extends far beyond the physical spatial dimension of not being able to find one's way home or stumbling about a dark room, also featuring strongly in the subject's psychological dimension of identity construction: "Disorientation is a way of describing the feelings that gather when we lose our sense of who it is that we are" (ibid.).

Note

1 Parts of this chapter, which appear here in a revised and amplified form, have already been published in Schmidt di Friedberg M. (2007).

5 "On the origin of certain instincts"

Ice bars my way to cross
the Yellow River,
Snows from dark skies to climb the T'ai-hang mountains!
(Hard is the journey, Hard is the journey,
o many turnings,
And now where am I?)

Li Po (701–762) 'Hard is the journey'

Disorientation of bodies that do not quite belong[1]

This chapter addresses the relationship between disorientation and the construction of the "other" in relation to space.[2] As we have seen, the word "orientation" is etymologically derived from "orient", meaning east, while the concept of East is practically a European invention, associated with women, the irrational, and the exotic (Said, 1978). The debate about the embodied subject, and the universal norms with which this subject is expected to conform, extends beyond gender relations and sexual orientation to include other forms of diversity. In the words of Puwar: "Some bodies are deemed as having the right to belong, while others are marked out as trespassers, who are, in accordance with how both spaces and bodies are imagined (politically, historically and conceptually), circumscribed as being 'out of place'. Not being the somatic norm, they are space invaders" (2004, p. 8). The "less-than-white" feature prominently among the "space invaders" identified by Puwar, who argues that race is a key category used to analyze "others". The colonial stereotype of racial superiority is founded, amongst other elements, on "the schizophrenic dualism" of "negro-white" (Fanon, 1963). Dualisms of this kind – man–woman, reason–sentiment, human–animal, body–soul, east–west[3] – provide a rationale for classifying the "other" as inferior, within the "supremacy of the 'White Man'" paradigm constructed by colonialism. "Black bodies" find themselves in spaces in which they are not entitled to be, and under the scrutiny of the white gaze, which constantly reminds them that they are "out of place".[4] Disorientation is the characteristic condition of these "dissonant bodies" and "one of the processes that bring to the fore the space-invader status of racialized bodies in privileged occupational positions. It is revealing of how specific bodies have been constructed out of the imagination of authority. Disorientation does also occur along the lines of gender. But it is not as acute as it is with 'race'" (Puwar, 2004, p. 43). The arrival of different bodies, as when migrants attain prestigious positions within white society, can engender both fear and disbelief:

> The dissonance caused by the arrival of women and racialised minorities in privileged occupational spaces unleashes shock and surprise. Their entry causes disorientation and terror. The threat they are seen to pose amplifies their presence. As "space invaders" they represent a potential organisational terror. They are thus highly visible bodies that by their mere presence invite suspicion and surveillance.
>
> (ibid., p. 54)

From the phenomenological perspective brought to bear by Sara Ahmed, our "corporeal schema" is racialized following a logic of exclusion: "If the world is made white, then the body at home is one that can inhabit whiteness" (Ahmed, 2006, p. 6). The "other" comes across as irrational, instinctive, and emotionally driven, in relation to a social model based on the undisputed supremacy of white, and male, rationality:

> At times this Manicheism goes to its logical conclusion and dehumanizes the native, or to speak plainly, it turns him into an animal. In fact, the terms the settler uses when he mentions the native are zoological terms. He speaks of the yellow man's reptilian motions, of the stink of the native quarter, of breeding swarms, of foulness, of spawn, of gesticulations. [...] that mob without beginning or end, those children who seem to belong to nobody, that laziness stretched out in the sun, that vegetative rhythm of life – all this forms part of the colonial vocabulary.
>
> (Fanon, 1963, pp. 42–43)

In narrating his experience as a black intellectual in France, Fanon identified his encounter with the white gaze as the start of his disorientation:

> And then the occasion arose when I had to meet the white man's eyes. An unfamiliar weight burdened me. The real world challenged my claims. In the white world the man of color encounters difficulties in the development of his bodily schema. Consciousness of the body is solely a negating activity. It is a third-person consciousness. The body is surrounded by an atmosphere of certain uncertainty.
>
> (2008, p. 83)

Thus bodies are constructed by the white colonial gaze, which perpetually prescribes a single standard model for them to conform to. Fanon elaborates: "Below the corporeal schema I had sketched a historico-racial schema. The elements that I used had been provided for me not by 'residual sensations and perceptions primarily of a tactile, vestibular, kinesthetic, and visual character,' but by the other, the white man, who had woven me out of a thousand details, anecdotes, stories" (ibid., p. 84). This view is in turn interpreted by Ahmed who comments:

> For Fanon, racism "stops" black bodies inhabiting space by extending through objects and others; the familiarity of "the white world", as a world we know implicitly, "disorients" black bodies such that they cease to know where to find things – reduced as they are to things among things. [...] The disorientation affected by racism diminishes capacities for action.
>
> (2006, p. 111)

Phenomenology, according to Ahmed, can help to unpack the construction of race and racialization and explore how the invention of race "as if it were 'in' bodies[,] shapes what bodies 'can do'" (ibid., p. 112). In the colonial era, the construction of bodily norms based on white supremacy took the form of specific behavioural rules in the social domain, education system, and scientific discourse. Racism is implicit in environmental determinism and social Darwinism. The "other", that is to say, the savage encountered by colonial "explorers", was framed as inevitably lacking in logical reasoning ability: "It would be easy to fill a volume with the evidence of excessive stupidity recorded by different travellers" (Lubbock, 1913, p. 565), wrote Sir John Lubbock, later Lord Avebury, a disciple of Darwin's. The "stupidity of savages" reported by the colonial authors was proved by the "natives'" supposed inability to achieve anything of worth, their poor mathematical skills, and lack of rational faculties. All of these negatives were documented in a vast interdisciplinary literature that on the one hand reflected the dominant culture of the era – marked by widespread and virtually unbounded prejudice towards the racial "other", and on the other established scientific theorizing as the only rational perspective on the world. In a corollary to the "stupid savage", the stereotype of the "difference" between "us" and "them" also underpinned the "good savage" theory, which proposed that indigenous peoples had been extraordinarily successful in adapting to their natural habitat, developing the capacity to survive in extreme conditions. The "savage" (and his female partner) were portrayed as magnificent animals, often physically superior to the "white man", who might enjoy lesser bodily strength but was always better endowed intellectually. In this theoretical take on white supremacy, orientation provided a key focus: "Members of primitive societies possess the instincts of animals. A savage displays exactly the same sense of smell and touch as a big cat. He is also endowed with the orientation instinct" (Thevenin and Coze, 1929, p. 70).

"Cet instinct qui sert à l'indigène de guide infaillible"[5]

For two centuries, the accounts of travellers and ethnographers conveyed an ambivalent image of the so-called "primitive" peoples, helping to frame these populations as distant from white supremacy: members of indigenous communities were portrayed as totally lacking in logic and reason, though often possessing wonderful and infallible senses and instincts (sight, hearing, sense of direction). Rousseau made an early contribution to the field:

> Savage man […] will have […] highly acute sight, hearing and smell. Such is the animal state in general, and, according to travelers' reports, such also is that of most savage peoples. Thus it is not surprising that the Hottentots of the Cape of Good Hope catch sight of vessels on the high seas with their naked eyes from as far away as do the Dutch with spyglasses, nor that the savages of America track the Spaniards by smell just as well as the best dogs could have done.
>
> (Rousseau, 2012, p. 71)

Pierre Jaccard, in an extensive 1932 study on sense of direction and orientation in humans, adopted a somewhat more critical stance on the fantastic tales of travellers, ethnographers, and pigeon-fanciers. Amongst other material, Jaccard cited the missionary Ch. W. Chatelain's account of his encounter with a blind native of the African region of Transvaal, who safely returned home after a lone journey of over 300 kilometres across dangerous desert territory with the sole assistance, according to the missionary, of: "that instinct which acts as the native's infallible guide" (Jaccard, 1932, p. 13). And similar accounts were arriving back from all parts of the "non-civilized" world: "On the other hand, the native Australian, just like the animal in whose material existence he partakes, is characterized by astonishing agility and exceptionally keen senses. He is endowed with the eyesight of a bird as well as a dog's sense of smell: he sniffs at clods of earth to find his way. Using touch alone, he can follow a trail by night" (Hardy, 1933, p. 36).

For decades, travellers to all regions of the colonial world (though often failing to specify the precise geographical locations visited) continued to report instances of the infallible instinct or sense of direction of indigenous peoples, while systematically ignoring numerous instances of loss of direction on the part of local guides. In his essay, Jaccard critically reviewed the leading theories on "sense of direction" and a range of examples of the so-called infallible sense of orientation of "non-civilized" peoples, pointing out that much of the travel lore was: "not just exaggerated but also inaccurate". Based on a review of written travel documents, he showed that the myth of the "good savage" had been constructed, at least in part, on the basis of incomplete observations or distorted reports of early European expeditions to unfamiliar regions. For over two hundred years, a flow of frequently undocumented and ever-more colourful narratives had continuously fuelled the stereotype of the (physical) superiority/(mental) inferiority of "savages". For instance, it was reported of the mysterious Kaoeboe people of Sumatra that: "Their sense of direction verges on the prodigious, given that the virgin forest, in which they live, is all but impenetrable, being 'without sunlight and without paths'" (ibid., p. 11). Still more significant is a citation from a letter written by a certain Dr. Kent-Monnet and published on 15 January 1924, in the *Mercure de France*: "In Africa, a most remarkable sense of orientation is displayed by the little Negro girls. Many colonials relate that, when lost in the desert, even when accompanied by (native) guides, they only managed to find their way again by taking children in their arms, who when placed at the four cardinal points, never chose the wrong direction" (ibid.).

Such testimonies, as vague as they were unreliable, contributed to the pseudoscientific and ideological creation of racist theory. Stephanie Chaffrai has explored the triangular relationship between the body, landscape, and power, concluding that: "Colonial discourse manipulates the body and the landscape in order to appropriate them" (2005, p. 1). Travel reports facilitate symbolic appropriation of colonial space and its inhabitants; behind the ambivalent construction of the "savage", lurks the ideology of colonial discourse. Gobineau, in 1853, reached the following conclusion about the "redskins": "Thus, the American nations only saw the light of civilization for one fleeting moment, and on a dark day at that. And now they have returned to their usual state: a sort of intellectual void, out of which nothing has the power to drag them barring physical death" (Gobineau, 1854, p. 525).

However, admiration for the physical capabilities and sense of orientation of "savages" (which serve to distance them from "us") predates both evolutionary theory and the work of Rousseau and Diderot: in fact, from the end of the fifteenth century onwards, the narratives of great travellers and historians – from Montaigne, to Cartier, to Tocqueville – had inspired the idyllic image of the "good savage", a myth that would later prove functional to colonialism by drawing an ever-sharper distinction between the "savage" – gifted with extraordinarily developed senses and instincts but unable to engage in abstract reasoning – and white supremacy. From Columbus onwards, it was the extraordinary American landscape and its inhabitants to fire the European imagination with the myth of the "Indian", "child of nature", and his superior physical abilities. In 1703, Baron de Lahontan described the aptitude displayed by Native Americans in the Great Lakes region for: "cutting straight across the forests for a hundred leagues without losing their way, or following the trail of a man or a beast over grass and leaves" (Lahontan, 1990, pp. 647–648). Jaccard compared the experiences of two Jesuit priests who kept voluminous memoirs on the lifestyles of Native American tribes in Canada: Joseph François Lafitau who lived with Canadian Iroquois communities for five years at the beginning of the 1700s, and Pierre François-Xavier Charlevoix who spent time in the Saint Lawrence region some decades later. Lafitau's record of the habits and customs of Native Americans inspired Rousseau's concept of "the state of nature". Nonetheless, Jaccard considered his descriptions to be accurate and reliable. Lafitau was a careful observer, and having noted the Iroquois' ability to orient themselves using the sun, stars, wind, animal tracks, and other natural elements and landmarks, emphasized above all their habit of training themselves in observation and memory skills: "It is not that they have better or keener eyesight than we do, it is the outcome of careful attention and long habit" (Lafitau, 1724, p. 245).

Charlevoix, in contrast, was held by Jaccard to be: "the main culprit for the long and gradual deformation of fact that was to culminate in absurd nineteenth century theories concerning an instinctive sense of direction in primitive peoples" (Jaccard, 1932, p. 131). Though positively regarded by a number of historians, Charlevoix was accused by Jaccard of indiscriminately plagiarizing and distorting the writings of Lafitau. With regard to sense of orientation, it was Charlevoix who first came up with the idea of the innate instinct and superior sensorial abilities of North American "Indians":

> The perfection of the savages' inner and outer senses is the first advantage that they have over us. […] Their imagination is prodigious; they only need to be in a place once to acquire an accurate notion of it that will never be erased. No matter how vast or how untrodden a forest, they can cross it without going astray.
>
> (Charlevoix, 1744, p. 304)

He spoke in a similar vein of the "savages'" seafaring skills: "They would travel for 30 or 40 leagues on the open sea without a compass, returning to the shore at the very place at which they had planned to land" (ibid., p. 304).

Lafitau had referred to a "quality that *seems* innate". For Charlevoix, this quality became an instinct: "These peoples possess an admirable talent, I would be prepared to say an instinct, for knowing if they have already been in a particular place" (Charlevoix, 1744, p. 239), which he explains in the following terms: "They are born with this talent; it is not the outcome of their observations or of great practice; children who have never yet left their own village, step as confidently as those with the most experience of travelling the country" (ibid., p. 304).

The instinct debate in *Nature*

Instinct as an explanation for human and non-human behaviours was a key theme within the nineteenth-century scientific debate. Is orientation an instinct or an ability developed through practice? In the period spanning the late nineteenth and mid-twentieth centuries, the debate honed in on the extraordinary, and often superior, orientation skills of "savages" and the question of whether this ability was based on an "instinct" shared with other animal species. Cuvier's fixist dogma of instinct, dominant in the early part of the nineteenth century, was challenged by Charles Darwin in *On the Origin of the Species* (1859). For the following 70 years, controversy raged over the issue of the orientation instinct and whether animals and "non-civilized" peoples possessed a sense of direction unknown to members of "civilized" societies. Morris comments: "As the 19th century scientists were converted to evolution, they were thus also convinced of racism. They were certain that the white race was superior to other races, and the reason for this superiority was to be found in Darwinian theory" (Morris, 1973, p. 159). The publication of Alfred Wallace's *Contributions to the Theory of Natural Selection* (1870) (criticized in *Nature* by Claparède) and Darwin's *Descent of Man* (1871) sparked off a heated controversy in the journal *Nature*, which lasted throughout the 1870s with contributions peaking in the year 1873:[6] the exchange of correspondence focused on the "Instinct Question" and mechanisms of perception (smell, sight, hearing, direction) in the different species.

The leaders of this debate – who alongside Wallace, one of the few nineteenth-century non-racists, and Darwin himself, included Edward Cox, George Henry Lewes, Henry Forde, and George J. Romanes – presented and analyzed a fascinating series of living creatures (insects, migratory birds, sheep, dogs, horses, humans) known to be able to find their way back to their point of origin, by means of a mysterious skill or instinct: "By the aid of the unknown power by which many animals (and savage men) can retain a true course" (Darwin, 1883, p. 359). Evolutionary theory did away with the distinct separation between human and animal proposed by the fixist school and focused on the concept of instinct. In reality, Darwin himself was cautious in his approach to instinct, recommending prudence in attributing: "to migratory animals any capacity in this respect which we do not ourselves possess; though certainly in them carried to a wonderful perfection" (Darwin, 1883, p. 357). He removed his chapter on instinct from *The Origin of the Species* and it was only published posthumously by Romanes as an appendix to his *Mental Evolution in Animals* (1883). Darwin also wrote a paper entitled "On the Origin of Certain Instincts", published in *Nature* on 3 April 1873. Both chapter and article dealt with the transmission of behavioural traits and patterns in insects and other animal species, especially the extraordinary capacity of living things to find their way home. Darwin did not neglect to mention cases of animals that lose their way, such as woodcocks, salmon, and large numbers of swallows, the last-mentioned frequently encountered by ships on the Atlantic. As the discussion continued, it was extended to cover all animal species, insects (Lubbock[7]), migratory birds and, in particular, homing pigeons (see § 5.5).

In addition to cases of animals, however, both of Darwin's texts on instinct mention the 1821 expedition of the explorer Ferdinand von Wrangell to Northern Siberia in the environs of the Bering Strait, thus contributing to the myth of a superior sense of orientation in "savages". In his account of the trip, Wrangell described the unfailing ability of local guides to find their way in the Arctic regions. The "savages" in this instance were Cossacks, that is to say, citizens of Russia, whom the author even mentions by name. Wrangell writes:

> At this stage of my narrative I will notice the remarkable skill with which our sledge-drivers preserved the direction of their course, whether in

> winding among large hummocks, or on the open unvaried field of snow, where there were no objects to direct the eye. They appeared to be guided by a kind of unerring instinct. This was especially the case with my Cossack driver, Sotnik Tatarinow, who had had a great experience in this skill.
>
> (Wrangell, 1841, p. 140)

The Russian explorer then went on to explain that the guides used parallel crack formations in the ice pack to orient themselves: “The inhabitants of the *tundras* […] know by experience at what angle they must cross the greater and the lesser waves of snow in order to arrive at their destination” (ibid., p. 141). Darwin however cited the English translation of Wrangell from memory (as he himself acknowledged in *Nature*) and by generalizing from the explorer’s account gave birth to the myth of an extraordinary sense of direction in the Arctic peoples (Jaccard, 1932): “To give one instance, the experienced navigator Wrangell expatiates with astonishment on the ‘unerring instinct’ of the natives of N. Siberia, by which they guided him through an intricate labyrinth of hummocks of ice with incessant changes of directions; while Wrangell himself ‘was watching the different turns compass in hand and trying to reason the true route, the native had always a perfect knowledge of it instinctively’” (Darwin, 1883, p. 357). According to Darwin, Wrangell: “… there describes the wonderful manner in which the natives kept a true course towards a particular spot. […] with no guide in the heavens or on the frozen sea. He states […] that he, an experienced surveyor, and using a compass, failed to do that which the savages easily effected” (Darwin, 1873, p. 418). Nonetheless, Darwin did not attribute this ability (“unconscious” on the part of the indigenous peoples) to a “sense” that “we” do not possess, but rather to eyesight and muscular movements.

While on the one hand the presumed orientation skills of “savages” were portrayed as instinctive, on the other they were held to be organized strictly around local reference points as opposed to using a rational system of coordinates. As claimed in a 1913 article by C.C. Trowbridge published in *Science*, it was in terms of method that the orientation of “civilized men” was thought to differ from that of other “living creatures of a lower order” who lack the faculty of abstract reasoning: the former, thanks to their education, followed an egocentric method, based on the four cardinal directions and use of the compass; the other creatures, birds, beasts, fish, insects, children (there was no reference to women in this article), as well as the “large proportion of mankind living in an uncivilized state”, used a domicentric method, guided by the point from which they set out and the location of their own homes. Trowbridge emphasized: “… the entirely different mental concept of civilized human beings, on the one hand, and of the other living creatures, on the other, relating to space on the earth’s surface. The former look outward toward the horizon, the latter look backwards toward their starting point” (Trowbridge, 1913, p. 890).

The superiority of the rational egocentric method was also upheld in an account of a trip to tropical South Africa by another disciple of Darwin's, Sir Francis Galton. According to Galton:

> The Damaras are bad guides considering that they are savages, and ought to have the instincts of locality strongly developed. On subsequent occasions, in retracing our routes over wide extents of country, it was a common amusement to try each other's recollection of the road by asking what would be the next object or next turn of the path that we should come to. But it is difficult to compare a European's idea of a country with that of these savages, as they look at it in such different ways, and have their attention attracted to such entirely different objects. A Damara never generalises: he has no one name for a river, but a different name for nearly every reach of it; [...] A Damara who knew the road perfectly from A to B, and again from B to C, would have no idea of a straight cut from A to C: he has no map of the country in his mind, but an infinity of local details. [...] The Damaras pick out their way step by step; they never dream of taking a course and keeping to it. All their observations are directed to spoors, sticks and stones, and they perpetually look down on the ground and not round about them.
>
> (Galton, 1853, p. 106)

Nonetheless, the rational superiority of the egocentric method, especially if: "wandering through any maze-like region, such as a primeval forest" might be responsible: "for man's confusion when attempting to find his way" (Trowbridge, 1913, p. 890), leading even a "civilized" man to lose his way in unknown territory.

Animal migration

The discussion on instinct in *Nature* brings us back to the "nature *vs.* nurture" controversy (see Chapter 4) and the key question as to whether the need and capacity for orientation are determined by biogenetic factors, that is to say, by congenital or hereditary conditions, or rely on culturally and experientially acquired skills. In the words of Stegmaier: "The form of orientation that is most evident to human beings is the independent spatial behaviour of living things. This ranges from 'orientation movements' in response to immediate stimuli, to periodic 'migrations' between habitats" (Stegmaier, 2008, p. 35). Orientation, in both humans and non-humans, involves cognitive activity that is more than simple sensory perception: "The ability to move is one of the most important characteristics of living things. Migration in all its often astonishing variety is the most elaborate expression of that ability, and its prevalence in so great a diversity of organisms demonstrates that it has been repeatedly favored by natural selection over alternative lifestyles" (Milner-Gulland et al., 2011, p. 413).

Now, orientation in living organisms is not a matter of purely mechanical reflexes, or automatic behaviours, but requires the processing of spatial information. The suggestion, long mooted by scholars (see the debate in *Nature*), that innate abilities or what has been defined as a special "sixth sense" (see above) play at least a part in orientation, prompts us to observe the fascinating forms of movement displayed by other living species and the complex dynamics underpinning them. Modern biology has shown plants to have orientation systems enabling them to effect movements – known as tropisms – in response to the force of gravity (geotropisms) and other external stimuli (light, temperature, humidity, chemical factors, etc.): "Some plants can only move to the right, others only to the left, confirming the assumptions of eighteenth century philosophical thinking about orientation" (Stegmaier, 2008, p. 36).[8] In the case of humans, enquiry into orientation necessarily involves asking whether and how innate abilities are enhanced or undermined by culture and practice, a question that extends to our bodies' relationship with technology and the hypothesis that we may be progressively losing our "natural" capacity to orient ourselves. More specifically, it has been speculated that members of technologically advanced societies – who have enjoyed the use of the compass for centuries and currently have access to more sophisticated navigation systems (e.g., the GPS and mobile phone devices referred to as "artificial prostheses" by Thrift, 2008) – are no longer motivated to observe natural phenomena, and therefore at risk of losing their "natural" orientation systems (see Chapter 10).

In free-moving organisms, orientation is linked with having a sense of balance. In higher animals, balance is governed by the vestibular labyrinth, which registers the body's movements in the different directions. Thus, modern biology provides scientific confirmation of Kant's and Husserl's early hypothesis that the body plays a key role in orientation, as well as tying in with the *analysis situs* of nineteenth-century mathematics (Poincaré, 1895). As affirmed by Ernst Mach in his historic lecture "On Sensations of Orientation" (1897) concerning the connection between balance and orientation, the vestibular apparatus in the human and non-human ear functions as an organ of balance and supports orientation, which does not have a specific dedicated organ of its own. Alterations in the vestibular system can cause loss of balance or vertigo.[9] Meanwhile, different animals display a wide diversity of efficient orientation systems, which includes forms of vectorial navigation in migratory birds, drawing on an inbuilt geomagnetic compass (e.g., snails), or using a "solar compass", as in the case of homing pigeons who can get lost when the weather is cloudy.

The extraordinarily diverse biological phenomenon of migration has fascinated human beings since antiquity. The "disappearance" of certain birds during the cold season has been observed by countless scholars over the centuries, from Aristotle and Plinius to Linnaeus and a huge variety of explanations put forward. In Olaus Magnus' *History of the Northern Peoples* (1555), a woodcut shows fishermen drawing swallows from the bottom of a lake.

According to Magnus: "the swallows hide in the waters" (p. 240)[10] to hibernate, coming out again in spring. Today, despite major advances in scientific biology and a growing database of detailed case-studies concerning different animal species: "there is still a dearth of underlying ecological and evolutionary theory, and a particular lack of general quantitative theory" (Milner-Gulland et al., 2011, p. 1). Overall however, research on the migratory behaviour of animals and plants (Dingle, 1996) increasingly suggests that migration is a highly complex phenomenon, in which not one but many factors come into play.

The long-distance navigation abilities of navigating birds and the extraordinary skills of homing pigeons, which have been used for hundreds of years both as a means of communication and in sporting contests, have yet to be fully explained: "Finally, the ability of true navigation found in birds remains one of the most enduring mysteries in biology" (Gauthreaux, 1980, p. 363). The *Nature* debate on instinct among Darwin, Wallace, and followers featured many letters and comments on the so-called "homing instinct", based on cases of animals that had, or had not, found their way home. Historically, there has also been much discussion about the abilities of carrier pigeons, a frequent object of study given their extensive use for military and sporting purposes. In the opinion of a Capt. Renaud, head of the French Army's carrier pigeon service, pigeons and other birds and animals endowed with the homing instinct display a "sixth sense" for orientation. As he declared before the French *Academie des Sciences*, citing evidence from numerous field experiments: "The instinct of distant orientation is that faculty which animals possess of taking the back track or retracing a way already taken. The sixth sense is apparently independent of the external impressions which affect other five senses" (Renaud, 1898).[11] In addition to a hypothetical "sixth sense", from Renaud's day to ours, many other abilities and theories have been invoked to explain how birds orient themselves during migration, yet: "After more than 50 years of intensive study, how homing and migratory birds navigate over great distances remains an open question" (Hagstrum, 2013, p. 687).

Nonetheless, the most recent research suggests that animal migration is underpinned by extremely complex systems that do not depend on any one factor but are driven by a variety of phenomena. In the words of Able: "Recent work with both migrant bird species and homing pigeons has revealed greater complexity than was previously expected: no single cue system or mechanism is likely to explain bird navigation. Birds have access to and apparently make use of a variety of sources of orientational information" (Able, in Gauthreaux, 1980, p. 285). The orientation mechanisms identified to date include: those based on visual data such as perceptions of light and observation of the movement of the sun, moon, and stars; the gradients in the geomagnetic field; models of the polarization of light; and the creation of mental maps (Kramer, 1959, Griffith, et al., 1989, Emlen, 1975, Wiltschko and Wiltschko, 1988, Muheim et al., 2006, Hagstrum, 2013). In homing pigeons, Able affirms that: "the map component of navigation has remained frustratingly elusive. [...] Although such pigeons utilize the sun compass in the determination of initial flight bearings [...] the navigational component of the homing process may well be based largely or entirely on nonvisual information" (Able, in Gauthreaux, 1980, p. 352). Many animals draw on a path integration system, which involves the ability to compute their current position as a function of their speed and acceleration, among other components. One of the most fascinating examples of orientation is the behaviour of the desert ant (genus *Cataglyphus*), studied by Wehner and colleagues (Wehner and Wehner, 1990), which is able to find its way in the desert by using the sun as an azimuthal coordinate. In addition to visual cues, orientation may also be based on the earth's magnetic field; currents and wave motion for aquatic animals; wind direction for birds, insects, and bats; acoustic cues, such as natural infrasound; and olfactory stimuli.

Despite migratory animals' extraordinary capacity for navigation, often over extremely long distances, many cases of disorientation and loss of the ability to locate reference points have been identified. Migratory birds, just like human airplane pilots,[12] can suffer from vertigo and spatial disorientation, particularly when flying at low altitudes by night. In both cases: "Studies have shown that disorientation is, almost without exception, the result of normal psychophysiological processes associated with certain characteristics of flight and of the pilot aloft [...]. Therefore, disorientation could be considered to be normal in the sense that it is a perceptual process correlated with the sense organs functioning normally in an abnormal environment" (Herbert, 1970, p. 402). Adverse weather conditions such as cloud cover, rainfall, and lightening, as well as manmade physical obstacles, may all lead to a loss of orientation. The scientific literature and news programmes abound with documented cases of birds that have crashed into antennae or electric wires and fallen to the ground in a state of disorientation. A phenomenon explained by Herbert based on his study of Blackburnian Warblers: "The bird is attempting to correlate compounding sensual information with visual information. As will be noted, it is the birds' inability to analyse the visual and sensory cues that is the basic cause of spatial disorientation. Unfortunately spatial disorientation, when flying, is a phenomenon which occurs with lightning rapidity" (Herbert, 1970, p. 416).

Many migratory bird species use the earth's magnetic field to orient themselves, although it is not yet clear how they detect it. Studies on homing pigeons that got lost during races in the US have shown that these birds rely on low-frequency electromagnetic waves to recognize the places they start out from. Under adverse atmospheric conditions, the earth and the winds can deviate ultrasound waves, making it impossible for the pigeons to recognize the site of their release and leading them astray (Hagstrum, 2013). Similarly, studies on European robins conducted at the University of Oldenburg show that low frequency electromagnetic waves produced by ordinary electrical appliances can prevent these birds from using the terrestrial magnetic field and therefore reduce their ability to orient themselves during migration (Engels et al., 2014). The potentially negative impact of electromagnetic activity generated by human activity and the possible inversion of the earth's magnetic field on the migratory movements of many species is still under study. Given that dolphins and whales rely on sound to navigate correctly, the phenomenon of mass cetacean stranding may be linked to geomagnetic distortions or to humans using sonar or other instruments with a high acoustic impact for military purposes or oil exploration. Another form of disorientation is linked with the phenomenon of "accidental displacement", a category of movement in migratory organisms that is not determined by the habitual behaviour of the species, but rather by unforeseeable circumstances leading animals and plants to colonize new habitats: "The ornithological literature is replete with examples of birds being carried beyond or outside their usual ranges by storms or periods of prevailing winds in the 'wrong' direction [...] Many terrestrial insects may also be blown off course" (Dingle, 1996, p. 17). Finally, it is becoming increasingly urgent to study the impact of climate change on migratory phenomena. For climate change, along with the loss of many habitats and the disruption of others, altered arrival and departure times in response to climatic variables such as temperature and rainfall, and variations in the availability of resources along the migratory path, can contribute to interrupting or modifying the migratory patterns of many species (Ahola et al. 2004, Wilcove, 2008, Milner-Gulland et al., 2011). Manmade structures and obstacles, ranging from barbed wire along roads and railways in the case of terrestrial mammals; dams, banks and other barriers in the case of salmon and other fish species; roadways and other obstructions in the case of frogs, salamanders and other amphibians; and skyscrapers and cities in the case of birds; in addition to the destruction of forests and wetlands, pollution, hunting, and over-exploitation, may all lead to a general decline in migratory behaviour as well as diminished animal populations. In Wilcove's view: "Simply stated, the phenomenon of migration is disappearing around the world" (2008, p. 5).

Notes

1 Puwar (2004, p. 144).
2 Part of this chapter, presented here in a revised and extended form, was previously published in Schmidt di Friedberg (2014).
3 Cfr. Intersecting Western dualisms, adapted from Plumwood (1993) and Gaard (1997).
4 In the chapter *Disorientation in Academia,* Puwar cites the colourful example of Lévi-Strauss, who while carrying out research for his *Elementary Structures of Kinship* in the reading room of New York Public Library, was struck to see an Indian in a feathered headdress "taking notes with a Parker pen": "The arrival of a feathered Indian with a Parker pen (an instrument of technology that has written the world into being) into the reading-room (a place from where the world is contemplated) is discomforting not merely because the analytical categories of this scholar are not sophisticated enough to fit the image, but, more importantly,

because the very identity of the intellectual as sovereign knower of the world is called into question" (Puwar, 2004, p. 45).

5 "That instinct which acts as the native's infallible guide".

6 Cfr. "Man and Natural Selection", Nov. 3, 1870; "The Difficulties of Natural Selection", Dec. 1, 1870, Dec. 8, 1870; "The Intelligence and Perfectibility of Animals", Jan. 5, 1871; "Modern Application of the Doctrine of Natural Selection, Feb. 13, 1873; "External Perception in Horses", "External Perception in Dogs","Mr. Wallace on Instinct", Feb. 13, 1873, March 6, 1873; "Inherited Feeling" Feb. 20, 1873; "Perception and Instinct in the Lower Animals", March 20, 1873, May 22, 1873, August 14, 1873; "The Sense of Smell in Animals", Mar. 13, 1873; "Perception and in the Lower Animals", March 27, 1873; "Origin of Certain Instincts", April 3, 1873; "Instinct", April 10, April 17, April 24, May 1, 1873; "The 'Instinct' Question", May 22, 1873; "The Migration of Birds", Sept. 24, 1874, October 29, 1874; "Intellect in Brutes", May 1, 1879, June 16, 1879.

7 See Lubbock (1882).

8 On right and left, cfr. Vilma Frisch, *Links and Rechts in Wissenschaft and Leben*, Stuttgart, 1964, cit. Stegmaier (2008, p. 36, footnote 3).

9 See also Berthoz (1997).

10 Magnus (1996, Book 19, Chapter 29, vol. 3, pp. 980–981).

11 Published: 13 February 1898. Copyright © The New York Times.

12 See Steyerl (2012).

6 Different spatial abilities

A phenomenological approach to orientation disorders

In this chapter, I examine a range of scenarios in which individuals can lose their "sense of orientation" or experience modified perceptions of space as a result of specific illnesses, brain damage, or the use of alcohol and drugs. I do not aim to provide exhaustive coverage of such conditions nor do I intend to dwell on their medical-scientific aspects, as this would fall beyond the scope of the book. Rather, I once again place the human subject at the centre of my discourse, exploring the spatial experience of disorientation from the insider perspective of the body-subject. My own interest in disorientation was originally prompted by my personal history as a long-term Type-1 diabetes sufferer, which led me to experience a complete loss of orientation during episodes of severe hypoglycaemia. In that condition, I recall wandering about in a confused state, with the perception that my thought processes were slowing down as I began to hallucinate. Nonetheless, in my memory of them, these episodes were never entirely unpleasant or terrifying. I would ramble about aimlessly, ignoring a vague recollection that there was something I urgently needed to do (reach my hand into my pocket to pull out a sweet), while succumbing to the appeal of the mental labyrinths that were opening up before me, giving me access (at least in my imagination) to the meanders of other spaces and dimensions.

The theme of illness once again leads us to invoke phenomenological and post-phenomenological approaches and, in particular, to revisit the work of Merleau-Ponty. From this author's perspective, in conditions of illness, the complex relations among body, mind, and space as we "live" them, as well as the distinction between the objective world and the lived world are called into question. In the sick or disabled body, everyday actions such as constructing a mental map or locating landmarks are never taken for granted. No longer situated in the unconscious sphere of "normality", they are now the outcome of a huge mental effort that brings all states of dysfunction to the fore. In the chapter of his *Phenomenology of Perception* entitled "Space", Merleau-Ponty compared his own concept of space with those of Berkeley, Descartes, and Kant. He disagreed with the notion of a single, "universal" space proposed by

Kant in his *Transcendental Aesthetic*, and defined as pure perception rather than the outcome of relations among things; instead, he himself advocated the idea of a multiple and subjective space: "There are as many spaces as there are distinct spatial experiences" (2005, p. 34).

In attempting to explain the relationship between space and the body-subject, Merleau-Ponty appealed to the experiments of Gestalt psychologists George Stratton and Max Wertheimer on the effects of distorting perceptions of space by inverting the images received by the retina: when special glasses were used to turn the experimenter's view of what was in front of him upside down, it proved possible to partly adapt to this altered, non-normal, mode of vision. Merleau-Ponty also drew on accounts of patients with brain damage or affected by body schemata disorders, who have lost the ability to correctly interpret the functioning of their own body, as occurs in individuals who perceive a phantom limb:[1] "It is precisely when my customary world arouses in me habitual intentions that I can no longer, if I have lost a limb, be effectively drawn into it, and the utilizable objects [...] appeal to a hand which I no longer have" (ibid., p. 95). Thus, space may be conceptualized as a subjective experience that is linked to the body. In keeping with this perspective, Merleau-Ponty explored psychiatrist Eugéne Minkowski's[2] accounts of how his schizophrenic patients personally experienced the world and space:

> The shrinkage of lived space, which leaves no margin to the patient, leaves no room for chance. Like space, causality, before being a relation between objects, is based on my relation to things. [...] Clear space, that impartial space in which all objects are equally important and enjoy the same right to existence, is not only surrounded, but also thoroughly permeated by another spatiality thrown into relief by morbid deviations from the normal.
> (ibid., p. 334)

In all the disorders examined by Merleau-Ponty, similarly to in dreams, myth, and the darkness of the night, disorientation appears to be the outcome of experiencing "pure depth": "A depth without foreground or background, without surface and without any distance separating it from me" (ibid., p. 330). Indeed, disorientation might be defined as an intrinsic dimension of pure depth that instantly comes into play when there is a breakdown in the relationship between the body and space and subjects lose the means of finding their way around a world that has come to seem alien and unrecognizable. Merleau-Ponty argued that:

> [w]hat protects the sane man against delirium or hallucination, is not his critical powers, but the structure of his space: objects remain before him, keeping their distance [...] What brings about both hallucinations and myths is a shrinkage in the space directly experienced, a rooting of things in our body, the overwhelming proximity of the object, the oneness of man and the world, which is, not indeed abolished, but repressed by

> everyday perception or by objective thought, and which philosophical consciousness rediscovers.
>
> (ibid., p. 339)

The theme of our relationship with space under conditions of illness was revisited by Heidegger in *Building Dwelling Thinking*, specifically in relation to the concept of dwelling: "Indeed, the loss of rapport with things that occurs in states of depression would be wholly impossible if even such a state were not still what it is as a human state: that is, a staying with things" (Heidegger, 1971, p. 155). Other authors (Yardley, Stiegler, Kvale) have contended that disorientation is an increasingly widespread phenomenon in our contemporary era, at both the cognitive and metaphorical levels: "The discursive and material aspects of this progressive transformation of the modern world are inextricably linked" (Yardley, 2002, p. 125). This may be due to the accelerating pace of change in our lifestyles and our lack of a sense of belonging to the places-non-places of post-modernity:[3] "The labyrinthine and self-reflecting postmodern space, as in the new hotels, makes it difficult for the body to be oriented in space. This post-modern space may be seen as a symbol of the inability of consciousness to grasp the larger, global, multinational and decentralized communication networks of a postmodern society" (Kvale, 1992, p. 39). Another cause suggested by Yardley for a generalized sense of disorientation is the fact that we now habitually avail of rapid means of transport, while at the same time our lifestyle is more sedentary than ever before:

> The prolonged mild disorientation which commonly follows vestibular disorder in Western populations may therefore be partly a product of our relatively sedentary modern lifestyle; in former times the luxury of avoiding vigorous activity (and thereby indefinitely retarding adaptation) was not an option. Persistent vertigo may also be maintained by cultural conventions which discourage adults from engaging in the activities which children still naturally employ to develop their orientation capabilities, constantly hopping and dancing, swinging around lampposts, turning somersaults, or balancing on narrow walls.
>
> (Yardley, 2002, p. 119)

The work of Merleau-Ponty and phenomenology offer a tool for the philosophical interpretation of mental illnesses as they affect the subject's relationship with space. Accordingly, Joyce Davidson, in her research on mental health, anxiety disorders, and phobias, has taken a phenomenological approach to examining perceptions of space in persons with agoraphobia:

> So long as lived space is successfully maintained, we have a sense of freedom to move as we please (within the limitations of our physicality),

> an easy relation with an environment that invites and encourages our involvement. But, if our sense of space breaks down we may find ourselves facing a repellent, unwieldy environment that restricts our freedom of movement, and threatens our sense of wellbeing.
>
> (Davidson, 2000, pp. 645–646)

The body-subject is at the heart of any phenomenological account. Hence, we are not "ghosts in a machine" but embodied entities:

> We are not Cartesian subjects. There are no fixed and immutable boundaries between self and world. The self is a project that needs to be constantly renegotiated and the senses provide the medium of this dialectic and of our orientation in social space. The experience of panic can arrive "out of the blue" precisely because [the]very sensation carries within it the germ of a dream or depersonalisation [… that] takes place on the periphery of my being.
>
> (ibid., p. 655)

Merleau-Ponty provided a framework for investigating the relations between "objective space" (geometric, Cartesian) and "lived space" defined as: "The projection of our own spatial orientation on to the world, also referred to as subjective, existential, anthropological or phenomenal space" (ibid., pp. 643–644). In patients with mental illness, this framework allows us to build up an understanding of: "The fragility of our existence, into the complex interrelations between body, space and self, into what it feels like, and what it means to be alive" (ibid., p. 643). The phenomenological approach and non-representational theory encourage us to revisit cognitive reference models and our view of the relations between body and space, mind and environment:

> Until quite recently, the study of cognition was based upon the classical cognitive model of representation in which passive representation of the world is followed by inference. But this kind of approach has been increasingly replaced by the notions of the direct perceptions of the unfolding of action-in-context. The environment is no longer passive. Instead it becomes a manifold of possibilities in time – and perception becomes a modulating trajectory which describes how the world is and simultaneously prescribes a space of adaptive responses. The environment is, if you like, an extension of the mind – and significantly, vice versa.
>
> (Thrift and Dewsbury, 2000, p. 415)

Feelings of vertigo or dizziness and difficulty in judging distances are frequently reported by patients suffering from phobic disorders, while from a neurological and cognitive viewpoint no direct relationship has been identified between panic attacks and agoraphobia on the one hand, and possible dysfunctions of the vestibular apparatus on the other (Golledge, 1999). As we have seen, pathological disorientation is attributed to a series of physical and mental causes that lead to feelings of distress or panic in certain spatial contexts. This is yet another example of the dualistic thinking typical of Western culture: specifically, the subject is viewed as separate from the surrounding environment and this justifies explaining the illness in terms of psychic or somatic factors. Lucy Yardley has investigated the phenomenon of disorientation and dizziness from an "ecological-constructionist" perspective. The starting point of her analysis, again informed by the phenomenological paradigm, is the historical development of the dominant biomedical model and the undoubted success of medical science in controlling and preventing many forms of illness and physical dysfunction and in reducing mortality rates:[4] "These triumphs are in turn attributed to ever-expanding scientific knowledge about the mysteries of anatomy, physiology, biochemistry, and (most recently) molecular genetics, together with increasing technological capabilities relating to procedures such as diagnostic testing and surgery" (Yardley, 2002, p. 3).

Yardley argues that in this "objectivistic" vision of medicine – in contrast with the approach practised by ancient or non-Western healers – subjective factors, beliefs, values, and emotions are frequently overlooked or seen as getting in the way of "true" factual knowledge: "The result has been the emergence of a dominant biomedical model of health and illness which is based on 'biological reductionism' – in other words, a belief that bodily events are best explained exclusively in terms of objective physical processes, and that these must be distinguished from the psychological and social factors which might bias or distort how physiological events are perceived" (ibid., p. 4). Yardley therefore offers a radically alternative interpretation of disorientation; "challenging both the dualism of the mind-body dichotomy, and the individualistic emphasis on personal (physical or mental) deficiencies" (ibid., p. 117). She explores the relations between individual and environment, questioning a dualistic vision of disorientation that impacts on the organization of medicine and the healthcare system, prompting its partitioning into the areas of somatic illness and psychic illness, and perpetuating the "theoretical segregation" of these two domains (ibid., p. 114). Yet: "This is no simple psycho-somatic or somato-psychic relationship, but a profound and intricate interactive bond between the material and discursive aspects of disorientation, which creates and reproduces the condition and preconditions, meaning and implications of disorientation" (ibid., p. 119). From the point of view of those suffering from orientation disorders, this reductionist approach and the resulting care strategies frequently fail to address many aspects of their illness; nonetheless, they may perceive the somatic approach to explaining the causes of their disorientation as more reassuring, in that it: "not only provides social legitimation of their complaint and protection against allegations of moral or emotional weakness […], but also offers the hope and reassurance that the underlying cause of the dizziness can be identified and controlled by medical science" (ibid., p. 112). Furthermore, the illness can give rise to a pervasive sense of insecurity that calls into question core aspects of the individual's previous lifestyle. Patients are obliged to live with a new "way of being" that modifies their body, never leaves them, and is not always identifiable or understandable to the people around them. As Patti Lather explains in relation to the life stories of women with HIV/AIDS:

> With women living with HIV/AIDS, I first began to learn about getting lost in terms of what it means to not be in control and to try to figure out a life, given that. Perhaps more precisely, I learned about getting lost from trying to simultaneously produce and theorize a book about these women. There I put myself in an awkward position that was not so much about losing oneself in knowledge as about knowledge that loses itself in the necessary blind spots of understanding.
>
> (Lather, 2007, p. VII)

Drugs, hallucinations, tripping

The distortions in spatial perception investigated by Merleau-Ponty not only concerned mental disorders (schizophrenia), but also "extreme states of consciousness", including the effects of mescalin:

> Under mescalin it happens that approaching objects appear to grow smaller. [...] The walls of the room are 150 yards apart, and beyond the walls is merely an empty vastness. The stretched-out hand is as high as the wall, and external space and bodily space are divorced from each other to the extent that the subject has the impression of eating "from one dimension to the other". Sometimes motion is no longer seen, and people seem to be transported magically from one place to another. The subject is alone and forlorn in empty space, he complains that all he can see clearly is the space between things, and that this space is empty.
>
> (Merleau-Ponty, 2005, p. 328)

Since ancient times, experiencing altered states of consciousness – ecstasy, hallucinations, illuminations, and dreams – has been viewed in many cultures as a privileged condition to be attained by means of meditation, fasting, and drugs. Throughout human history, the consumption of hallucinogenic substances has been both a collective and a religious affair that has fulfilled therapeutic and ritual functions.[5] In contemporary society, the use of consciousness-altering drugs has become a mass phenomenon that is no longer linked to collective rituals but is practised for recreational, hedonistic, and escapist purposes. Using drugs is one of the possible responses to the sense of existential disorientation and crisis permeating the contemporary world and a means of evading reality: "Basically, the farther we go, the more the question of drugs seems inseparable not only from such tremendous questions as 'the concept', 'reason', 'truth', 'memory', 'work', and so forth, but inseparable also from the emergency rooms where all these things appear to gather symptomatically" (Derrida, 2012, p. 35).

The use of "psychotropic" substances – a term introduced in the 1950s, which is derived from the Greek for "turning the mind or soul" and refers to substances that affect mental functioning and behaviour – has been explored across a broad interdisciplinary literature. Jacques Derrida noted that: "'Drugs' is both a word and a concept [...] Already one must conclude that the concept of drugs is not a scientific concept, but is rather instituted on the basis of moral or political evaluations: it carries in itself both norm and prohibition, allowing no possibility of description or certification – it is a decree, a buzzword (*mot d'ordre*)" (ibid., p. 20). Artists, musicians, philosophers, and writers, including Samuel T. Coleridge, Thomas De Quincey, Robert L. Stevenson, Charles Baudelaire, Aldous Huxley, Gottfried Benn, Walter Benjamin, Jean-Paul Sartre, Antonin Artaud, Allen Ginsberg, Jack Kerouac, William S. Burroughs, Philip Dick, and Stephen King, to mention but a few,[6] have drawn inspiration from, and often provided detailed descriptions of, their own experience of consuming plant-based or synthetic drugs. It is not among the aims of this book to enter into the broader debate on the use of drugs; rather in the current section I touch on how the use of psychotropic substances can affect subjects' spatial experience, perceptions, and orientation,[7] starting from the encounter with the lotus-eaters in the Odyssey, when Ulysses bound up his sailors under the rowers' benches for fear that eating the flowers would make them forget the way home: "According to the narratives, natural hallucinogens appear also as 'confusing substances'. They have the capacity to disconnect the individual from reality. One's senses and feelings are disrupted to the point of entering a world that has lost its way. The individuals find themselves in a virtual space, dreamlike or absurd" (Hunt et al., 2011, p. 154).

Hallucinations[8] or "a unique and special category of consciousness and mental life" (Sachs, 2012, p. XIII) may be caused by a number of different illnesses (hypoglycaemia, delirium, fever, epilepsy, schizophrenia, etc.) and can be highly diverse in nature. Altered perceptions of space and movement are key elements of the hallucinatory experience. Spatial metaphors (for example, getting high, spacing out, or being far out) liken the experience of using drugs to that of making a journey. Linsday M. Banco in *Travel and Drugs in Twentieth-century Literature* (2010) used the concept of tripping to analyze the complex relationship between drugs and travel among twentieth-century writers: "More than just a generic definition, 'tripping' – to travel and to intoxicate, to trip and to trip – offers a map of the complex and unstable ways different texts represent one thematic in order to understand the other" (Banco, 2010, p. 3). Consuming "hallucinogenic" substances with the power to produce radically altered perception and extreme states of consciousness (mescalin, cannabis, psilocybin, ayahuasca, LSD, etc.) is reported, among other effects, to give subjects the illusion that they have access to other worlds.

They may feel as though they are being transported into new dimensions of space. Familiar places become strange new environments, often populated by animals, fantastic beings, and geometric shapes, heightening subjects' feelings of disorientation. The ancient tradition of Ayahuasca or Yagè, practiced in much of the Amazon rainforest, which involves consuming certain plants for medical-religious purposes under the guidance of shamans, is associated with potent hallucinations and visions. In the *Yage Letters* (1963), defined by Rod Phillips as "America's first psychedelic travel narrative" (Phillips, 2000, p. 126), William Burroughs and Allen Ginsberg told of their encounter with Ayahuasca in 1953: "Yage is space time travel. The room seems to shake and vibrate with motion" (Burroughs and Ginsberg, 2006, p. 50). The hallucinations caused by drinking the brew are predominantly visual in nature:

> [The] designs function as paths to be followed by the eye-soul on its way away from and back to normal space and perception. When asked about the role of design in vision, one ayahuasca specialist told me, "One should always stay inside the design in order to not get lost." [...] In their quality of paths, these designs describe and map known space, while suggesting an infinite extension into unknown territories yet to be explored.
>
> (Shannon, 2002, p. 347)

Walter Benjamin, on his part, has left us a detailed description of his wanderings about Marseilles in 1928 and his altered perceptions of space after consuming a preparation of hashish: "Now the hashish eater's demands on time and space come in force. As is known, these are absolutely regal. Versailles, for one who has eaten hashish, is not large, or eternity too long. Against the background of these immense dimensions of inner experience, of absolute duration and immeasurable space, a wonderful, beatific humor dwells all the more fondly on the contingencies of the world of space and time" (Benjamin, 2006, p. 49). With the modification of the dimensions of space and distance came a general sense of disorientation:

> You start to play games with spaces in general. You start to experience seductions with your sense of orientation. In a waking state, we are familiar with the sort of unpleasant shift we can voluntarily provoke when we are traveling in the back of a train at night and imagine we are sitting in the front – or vice versa. When this is translated from a dynamic situation into a static one, as here, it can be thought of as a seduction.
>
> (Benjamin, 2006, p. 23)

Aldous Huxley, like Burroughs and Ginsberg, represented the counterculture of the post-war period, "the first generation of writers to become celebrities of the modern mass media and to have suffered the wholesale displacement of the real by the image" (Harris, 2003, p. 23). In *The Doors of Perception*

(1954) and *Heaven and Hell* (1956), he gave an account of the states of consciousness at the "antipodes of the mind" that had characterized his personal experience with mescalin. The concept of space and spatial relations featured strongly in Huxley's description of his "little tour" of Los Angeles in 1953 while under the effects of mescalin, during a visit to what was then considered "The World's Biggest Drug Store": "The field of vision was filled with brightly colored, constantly changing structures that seemed to be made of plastic or enameled tin. 'Cheap', I commented. 'Trivial. Like things in a five-and-ten'. And all this shoddiness existed in a closed, cramped universe. 'It's as though one were below decks in a ship', I said. 'A five-and-ten-cent ship'." (Huxley, 1954, p. 13). Nonetheless, Huxley emphasized that his modified perceptions of space mainly acted as a backdrop to an intense focus on immediate experience and the meanings attributed to it by his imagination:

> "What about spatial relationships?" the investigator inquired, as I was looking at the books. It was difficult to answer. True, the perspective looked rather odd, and the walls of the room no longer seemed to meet in right angles. But these were not the really important facts. The really important facts were that spatial relationships had ceased to matter very much and that my mind was perceiving the world in terms of other than spatial categories. At ordinary times the eye concerns itself with such problems as Where? – How far? How situated in relation to what? In the mescalin experience the implied questions to which the eye responds are of another order. [...] Space was still there; but it had lost its predominance. The mind was primarily concerned, not with measures and locations, but with being and meaning.
>
> (ibid., pp. 6–7)

Losing one's mind and spatial disorientation

A wide range of pathological conditions and disorders can cause us to "lose our minds"[9]. Disorientation, as we have seen, may be related to altered states of consciousness or determined by the external environment; it may involve inaccurate interpretations of the spatial aspects of visual experience or distortions in topographic memory: "Disorientation is not an internal state, like pain, but arises in the course of interactions between an individual and their environment" (Yardley, 1997, p. 117). Spatial disorientation can concern the perception of spatial relations between the observer and an object or among objects located in space, the orientation of external stimuli, and the recall of information or of how to perform basic mental operations in space (Benton, 1985).[10] Loss of orientation is linked to the functioning of different but interrelated areas of the brain, and to the body's complex systems of biochemical, genetic, psychological, and cellular information (Richter, in Baur, 2010). Within this network of relations, as we have seen, a key role in spatial navigation and memory is played by the hippocampus. It has been demonstrated that damage or lesions to the hippocampus or its partial removal can impair spatial learning and disrupt wayfinding and spatial memory. Nonetheless, Dudchenko continues to define the hippocampus as: "just one component of an interlinked system that provides orientation and recognition of the environment" (2010, p. 151).

In addition to forms of mental illness caused by brain damage or a variety of other pathological conditions, features of the environment can contribute to feelings of distress or panic in particularly disorientating spatial contexts: "Giddiness experienced when looking down from a great height generates fear and this, in turn, protects us from life-threatening situations. Limits are set on our orientation within space. [...] human beings are not made for moving horizons or great depths" (Richter, in Baur, 2010, p. 109). The height and depth limits to our perception of space can give rise to distressful sensations of vertigo, posing a risk in conditions that are extreme for the human body, such as flight: "The three-dimensional environment of flight is unfamiliar to the human body, creating sensory conflicts and illusions that make spatial orientation difficult, and sometimes impossible to achieve."[11] Spatial disorientation (SD) is recognized as a possible cause of accident in air pilots: "Flying an aircraft is a challenging activity, and exposes the pilot to many potential hazards. One of the most significant of these is spatial disorientation. Spatial disorientation is defined as the inability of a pilot to correctly interpret aircraft attitude, altitude or airspeed in relation to the Earth or other points of reference. It is a very common problem, and it has been estimated that the chance of a pilot experiencing SD during their career is in the order of 90 to 100 per cent" (Newman, 2007). Indeed, the statistics attribute 5–10 per cent of all air traffic accidents, both military and civilian, and as many as 90 per cent of fatal accidents, to pilot SD.[12] The phenomenon of in-flight spatial disorientation and errors in calculating the aircraft's current position can vary as a function of the type of operation to be carried out and the stage of the flight at which it is required. Adverse weather conditions, tiredness, stress, and the pilot's level of training are all factors that can contribute to SD. Currently, both the phenomenon itself and its incidence are being researched by the military and civil aviation authorities of many countries. Particular emphasis is laid on improving awareness of SD as a potential risk, and the adoption of other prevention measures including making adequate pre-flight preparation to minimize the risk of accidents.

Another environmental factor that can disorient our perceptions of the environment is noise. Exposure to high intensity or continuous noise can lead to hearing problems, vertigo, and disorientation in humans and non-humans. However, the total absence of noise can be equally disturbing, as has been demonstrated in the anechoic chamber of Orfield Laboratories[13] in South Minneapolis. The insulation system of an anechoic chamber absorbs all sounds and electromagnetic waves. Inside them, the only perceptible noise is that made by one's own body. For many people, experiencing such "deafening silence" causes disorientation and hallucinations and becomes intolerable within a short space of time. Generally speaking, different levels of spatial disorientation may be experienced and individuals' ability to move independently through space may be undermined to a greater or lesser extent. For Kevin Lynch getting lost in an unknown city is a wholly terrifying experience, only comparable to the disorientation of those who, as a result of brain damage:

> have lost the ability to organize their surroundings. They may be able to speak and think rationally, even to recognize objects without difficulty, but they cannot structure their images into any connected system. These men cannot find their own rooms again after leaving them, and must wander helplessly until conducted home [...] Locations normally identified by many objects in context may be recognizable only by virtue of some distinctive, separate symbol. One man recognizes a room by a small sign, another knows a street by the tram car numbers. If the symbols are tampered with, the man is lost.
>
> (Lynch, 1960, p. 125)

A shattered world

Technological advances (functional magnetic resonance imaging) and recent research have shown that orientation is a highly complex function, which is linked to an extensive neural network and in particular to the system formed by the hippocampus and the subcortical structures. Brain lesions and dysfunction of various kinds can impair the ability to navigate one's environment, causing "topographical disorientation". Alexsandr Luria (1902–1977), one of the founders of neuropsychology, dedicated his life to studying the functioning of memory and cognitive abilities in patients with brain damage. According to Oliver Sacks, Luria's key contribution was the notion that the core functions of mind and brain are not solely biologically based but also crucially depend on experience, culture, and our individual relations with the world around us and with space: "Even the most elemental functions of brain and mind were not wholly biological in nature but conditioned by the experiences, the interactions, the culture, of the individual" (Sacks, in Luria, 1972, p. VIII). Luria attempted to develop a "Romantic Science", as he himself referred to it, in opposition to classical science, the outcomes of which he claimed often included "the reduction of living reality with all its richness of detail to abstract schemas" (ibid., p. X). This romantic scientific approach informed the "neurological novel" *The Man with a Shattered World: The History of a Brain Wound* (1972, with a preface by Sacks), in which Luria presented the extraordinary account of a young student of mechanics, Lev A. Zasetsky, who served as a lieutenant in the Red Army during WWII operations on the Western front. Seriously wounded in 1943 by a German bullet that damaged the occipito-parietal region of his brain, and was never extracted for fear of causing further harm, Zasetsky lost much of his capacity for perception and memory, also suffering serious impairment to his visual field and other brain functions, and regressing to a childlike state in terms of his perceptions and personal memories. After a lengthy period of rehabilitation, Zasetsky returned to his native village, and sought – with Luria's assistance – to recover fragments of his lost brain function, keeping a detailed, epically long, and fascinating diary, in which for 25 years he reported on the successes and failures obtained through his efforts. Zasetsky felt like a "newborn": functions that he had previously taken for granted were now entirely lost, at times he was not even able to recognize the different parts of his own body, he had to "learn" again from scratch to make the most elementary kinds of movement, to read, and to perform the simplest of mathematical operations.

Nonetheless, because his frontal lobes were intact, he was fully aware of his own situation:

> But it was so hard to write. I'd get an idea of how to describe the moment I was wounded and the period right afterwards, when my illness began. [...] I'd try to clamp the words to the idea as much as I could. But what a torture it was. I always forgot what I wanted to write, what I had just been thinking the minute before.
>
> (Luria, 1972, pp. 79–80)

Affected by what would later be termed anterograde amnesia, following on his injury, Zasetsky was no longer able to create new memories. One of his impaired functions was precisely that of spatial perception. Formerly expert in geographical map-reading, which had been part both of his studies at the Polytechnic and of his experience in the military, Zasetsky was now unable to find his way around places that he had known since childhood, and to identify landmarks or read signs. He felt disoriented, often losing his way, and even failing outright to recognize sites that had previously been familiar to him:

> Suddenly the words right, left, back, forward, up, and down occurred to me, but they weren't any help since I didn't really understand what they meant. A minute later I also remembered the words south, north, east, and west. [...] I was lost. I didn't understand whether, north and south meant areas that were side by side or just the reverse. I even forgot what direction north or south indicated.
>
> (ibid., p. 53)

His disorientation did not improve with time, forcing him to remain close to his home, while a move to a new apartment made the situation worse: "During the first few days and weeks after we moved to the new place, I couldn't get used to it, couldn't get my bearings. [...] I can't remember how to get home. I not only forget where our apartment is, but even the name of my block. [...] I always have to carry a little notebook with me listing my address and apartment number, just in case I get lost" (ibid., p. 60). In the course of long conversations with Zasetsky, Luria observed that the serious brain damage caused by the bullet wound had played havoc with the patient's conceptualization of the world, and particularly with his concept of space: "Space 'made no sense'; he feared it, for it lacked stability" (ibid., p. 61). As these citations show, the voices of doctor and patient alternate throughout the book. After the preface by Oliver Sacks, in a brief introduction, Luria himself explains the basis on which this shared narrative was created: "It is not false modesty on my part to wish no credit for this book. The real author is its hero" (ibid., p. XIX). He collected a large volume of written material from Zasetsky, including his tattered war-time journals and other thicker notebooks covered in oilcloth: "There are almost three thousand pages in these. They represent twenty-five years of work this man devoted to describing the effects of a terrible brain injury" (ibid.). Luria's contribution was to guide us into the mind of a man who had lost his entire world, and being aware of this, fought desperately to get it back. The first title that Zasetsky proposed for his work was "I'll Fight On". As he himself put it: "It's depressing, having to start all over and make sense out of a world you've lost because of injury and illness, to get bits and pieces to add up to a coherent whole" (ibid., p. XXI).

Developmental Topographical Disorientation (DTD)

From a neurological perspective, disorientation is a theme that receives much attention, but concerning which many questions remained to be answered: "Human orientation in novel and familiar environments is a complex skill that can involve numerous different strategies. To date, a comprehensive account of how these strategies interrelate at the behavioural level has not been documented, impeding the development of elaborate systems neuroscience models of spatial orientation" (Arnold et al., 2013, p. 359). Iaria and colleagues have recently identified a new disorder in people (predominantly women) whose cognitive development is normal and who present no signs of brain damage, but yet are unable to spatially orient themselves in their everyday surroundings: "Topographical disorientation may occur as a development defect causing a lifelong disorder affecting daily activities (Iaria et al., 2008, p. 30). This illness, labelled Developmental Topographical Disorientation (DTD) is a rare neurological disorder that makes sufferers incapable of navigating any type of environment and of creating a mental map of their current location via the identification of landmarks. DTD is caused by "lack of the activation of the hippocampal complex and retrosplenial cortex while forming a cognitive map of the environment" (ibid.). The first observed case, patient Pt1, recalled experiencing panic from early childhood onwards due to getting lost on familiar routes such as going from home to school or even inside her own home. Other cases followed the first, allowing the researchers to broaden the scope of the study, and to investigate and formulate hypotheses about the new disorder from the structural, functional, and even genetic perspectives, as well as devising rehabilitation strategies for subjects whose orientation abilities are seriously impaired due to DTD.

As part of their programme to research and raise awareness of DTD, NeuroLab, a cognitive neuroscience research laboratory at the University of Calgary directed by Giuseppe Iaria, has developed the website gettinglost.ca, which offers easily accessible scientific information on human navigation and orientation. The site includes a forum for discussing personal experiences of disorientation, including DTD, and a testing platform on which users may participate in psychological research. In the words of the research team: "We want to gain a richer understanding of DTD, by observing behaviour, as well as the structure and function of the brain using magnetic resonance imaging (MRI), and characterize the cause of this disorder. [...] to help rehabilitate people that get lost easily, including people with DTD, the elderly, and individuals with acquired brain injuries or other neurological conditions. By using strategic interventions embedded in video games and real environments, we want to help people improve their ability to understand and navigate throughout their environment" (gettinglost.ca).

Aging and spatial cognition

Spatial cognition is a progressive process that modifies with age. An infant's first contact with its environment is mediated by its body, and this facilitates early experience and the development of intelligence. Gradually, the child acquires an image of its own body, a crucial reference point for the ongoing development of its ability to orient itself using the two key coordinates of time and space. Piaget, as early as 1948, had demonstrated the inseparability of motor and cognitive activity, proposing that children's cognitive development takes place over four stages, and is complete by about age 12 years. In the study *La représentation de l'espace chez l'enfant*, Piaget and B. Inhelder (1948) found that children gradually built up their spatial perception abilities, with their knowledge of the environment going through three increasingly complex stages of development, labelled egocentric, allocentric, and geocentric. At the third stage (geocentric), children construct a mental representation of space that is independent of themselves and of objects, and acquire the ability to make connections following general spatial laws, such as closeness, proximity, separateness, etc.

Towards adulthood, this representation system becomes more complex and flexible, being facilitated by language, while spatial knowledge is extended to include a wide range of environmental features (distal and proximal landmarks) and navigational strategies that the subject has learned to apply with increasing skill (Piccardi et al., 2011). From studies and experiments on orientation abilities in children, it has emerged that: "Children may reorient in a way that is qualitatively different from the adult" (Dudchenko, 2010, p. 107). Studies on how children construct their own personal spatial systems and on records collated by Search & Rescue Teams[14] of how children behave when they get lost, have explored young people's ability to observe features of the environment and its conformation, assessing whether in orienting themselves they mainly rely on geometric data or on landmarks. According to the work of Piaget and Inhelder, while children retain a predominantly egocentric perspective, they are unable to imagine the perspective of another person located at another point in space, and therefore have difficulty in using landmarks to navigate space: "Moving from these experimental observations to an account of why children get lost is, admittedly, a speculative exercise. However, it is possible that disoriented children may erroneously orient to the boundary features of an environment, and overlook potentially informative landmarks" (ibid., p. 107).

While, as we have seen, our spatial orientation system develops gradually over time, equipping us by adulthood with the ability to represent entire places in our minds: "in aging, it is possible that the fidelity of our spatial system degrades. Extrapolating from the results of place cell studies, our ability to reliably bring up the same cognitive map may be impaired. In addition, the flexibility of our map in terms of incorporating new information may be diminished. From a path integration perspective, our ability to track our movements and to integrate the distance and direction in which we have travelled may be less accurate than that of a young adult" (ibid., p. 254). A loss of orientation in the elderly and the ill was observed by Charles Darwin as early as 1873, leading him to hypothesize the existence of a brain region specialized in spatial activities:

> The manner in which the sense of direction is sometimes suddenly disarranged in very old or feeble persons, and the feeling of strong distress which, as I know, has been experienced by persons when they have suddenly found out that they have been proceeding in a wholly unexpected and wrong direction, leads to the suspicion that some part of the brain is specialised for the function of direction.
>
> (Darwin, 1873, p. 418)

With the passing of time, our body naturally changes and many of our physical and mental functions slow down and become less flexible. It is normal for older people to have greater difficulty in recalling names, places, and people and to take longer to process new information, including spatial data concerning position, direction, and orientation. Getting lost in unfamiliar surroundings may be one of the first signs of senile dementia, along with the continual searching for everyday belongings: "In fact, forgetting where things are – most likely 'mobile' objects such as keys or glasses – is one of the first clinical symptoms of dementia, illustrating the ecological significance of object-location memory research" (Postma et al., 2004, p. 144). Dementia is not a specific illness but a generic term used to describe a decline in cognitive functioning that may be underpinned by a range of different conditions which interfere with the fundamental activities of the brain. The areas that may be affected include memory, logic, language, behaviour, orientation, and personality. Today, over 35 million people in the world suffer from dementia, a figure predicted to double by 2050 (Feddersen and Lüdtke, 2014). Dementia may be more or less serious and can get progressively worse to the point of becoming irreversible, limiting sufferers' ability to cope with the demands of everyday life, and ultimately forcing them into residential care. It is caused by a range of neuro-degenerative disorders, the most widespread of which is the so-called Alzheimer's disease.

Alzheimer's disease

Alzheimer's disease was described for the first time in 1906 by the German psychiatrist and neuropathologist Alois Alzheimer. It is a form of degenerative dementia, which becomes progressively more invalidating and is associated with the formation in the brain of amyloid plaques and neurofibrillary tangles (Dudchenko, 2010, p. 223). Disorientation is often one of the first symptoms of Alzheimer's and, at Level 3 of the Global Deterioration Scale (GDS),[15] is ranked among the more serious forms of dementia that a subject can suffer (Reisberg et al., 1982); other symptoms include confusion, anxiety, and panic attacks: "In Alzheimer's disease, we know that there is damage in the brain regions which, in other mammals, contain head direction cells. This pathology may produce an inability to find one's way and, coupled with a disruption to the hippocampus and its inputs, a lack of familiarity with previous encountered locations" (Dudchenko, 2010, p. 254). From a spatial point of view, the main difficulties encountered by Alzheimer's patients concern recognizing familiar landmarks, even in their own homes, and wayfinding, likely on account of a deterioration in their cognitive ability to create a mental map: "Furthermore, the lack of good wayfinding abilities may be related to other negative symptoms of dementia, such as wandering (or restless walking): poor performance on simple wayfinding goals and global wayfinding strategies were the deficits most associated with wandering" (Marquardt, 2011, p. 24).

On the basis of a wide-ranging literature review, Gesine Marquardt (2011) identified three main causes for dementia-related spatial disorientation: memory deficit, visuospatial deficit, and specific changes in orientation strategies combined with a loss of ability to organize information. There have also been interdisciplinary studies of disorientation in persons with dementia, with a view to designing ad hoc environments for sufferers that facilitate orientation and reduce their sense of confusion and insecurity. It is critical for those with Alzheimer's disease to be effective in perceiving their living environment if they are to live independently. This can be brought about by creating highly specialized worlds:

> A final aside concerning the interplay between mind and environmental structure: Consider the case of patients with advanced Alzheimer's Disease. Many of these patients live surprisingly normal lives in the community, despite the fact that standard assessments of their capabilities suggest that many such patients should be incapable of surviving outside of special-care institutions. The key to such surprising successes, it seems, lies in the extent to which the individuals rely on highly structured environments, which they create and then inhabit. These environments may incorporate multiple reminding notices around the house and strict adherence to specific routines. One patient virtually lives on a couch in the center of her apartment, since this provides a vantage point from which she can visually access the location of whatever she needs – this really is a case of using the world as external memory.
>
> (Clark, 1997, p. 66)

Architecture, design, and urban planning can all play a role in designing new public spaces, residential care facilities, apartment complexes, and infrastructure that can offer dementia sufferers a certain degree of independence and a good quality of life:

> The design of the physical environment plays a major role in supporting the wayfinding abilities of people with dementia. The floor plan design of a nursing home in particular has a significant influence on residents' spatial orientation and wayfinding. Additional interventions such as signage, furnishing, lighting, and colors are additional supporting features but they cannot compensate for an adverse architectural design.
>
> (Marquardt, 2011, p. 75)

Given the progressive ageing of the population and the growing incidence of senile dementia, it would be of strategic value to plan entire city neighbourhoods to cater for this phenomenon, choosing architectural forms, street layout, and signage with a view to facilitating orientation and wayfinding. Residential care facilities might also be designed to offer easily identifiable points of reference and to guide orientation via signs, decorations, furnishings, the position and shape of entrances and the length of corridors. Personal belongings, photographs, and colours could also be helpful:

> Environmental interventions that promote way-finding can be implemented on two levels: the design of the floor plan typology and environmental cues, which comprise signage, furnishings, lighting, colors, etc. Even common materials that can be manipulated or used, such as groceries, water, and materials for handicrafts, can support residents' orientation.
>
> (ibid., p. 79)

As well as adapting and reinforcing visual cues, multisensorial stimuli should also be provided to attract patients' interest and enhance their wellbeing: "Environmental interventions to promote way-finding are best utilized when they are combined and address different senses" (ibid., p. 33). An example of this type of ad hoc planning is the village of De Hogeweyk (Weesp), near Amsterdam, created by the Dutch government with a view to optimizing the independence and opportunity for movement of its residents, who are dementia sufferers. This gated community is home to 160 senior citizens, divided into small groups on the basis of their lifestyle habits. The facilities and activities organized for the community (supermarket, theatre, green spaces, workshop) as well as the village's overall design are intended to maximize the residents' quality of life, encourage them to remain active and to reduce their reliance on medication vis-à-vis their counterparts in conventional residential care (Feddersen and Lüdtke, 2014).

Notes

1 "A phantom limb is the sensation that an amputated or missing limb (even an organ, like the appendix) is still attached to the body and is moving appropriately with other body parts" http://www.news-medical.net/health/What-is-a-Phantom-Limb.aspx. Accessed 9 April 2017. See also Berthoz (1997).
2 Minkowski (1932, p. 69).
3 "Nonplace is very different from a traditional humanistic conception of place as a location full of meaning to which people become profoundly attached. Nonplaces are marked by a lack of attachment, by constant circulation, communication, and consumption that act against developing social bonds and bonds between people and the world. These nonplaces are marked by a plethora of texts, screens, and signs which facilitate mediated relationships between people and places rather than direct ones" (Cresswell, 2009, p. 175). For discussion of non-places see also Augè (1995), Foucault (1986), Massey (1994), etc.
4 "Medicalising certain forms of disorientation is a gesture of bordering that bypasses the moment of identification of who is disoriented by what" (Martin and Rosello, 2016, p. 4).
5 This is no longer the case in the contemporary Western world: "Loss of control is not well countenanced in modern Western society, which is imbued with the ideal of individual autonomy, and the desire to rise above and subdue nature" (Yardley 2002, pp. 120–121). On the relationship between "mental illness" and exclusion see Foucault (1972).
6 The Beatles song *Lucy in the Sky with Diamonds* has sometimes been interpreted as referring to the use of LSD.
7 See Cohen (1965).
8 "In careful present-day usage, hallucination indicates a false appearance, in sensory form, hence seemingly external, but occasioned by an internal condition of the mind, the central suggestion of the term being its subjectivity and groundlessness" (La Barre, 1975).
9 The pathological causes of spatial disorientation include conditions affecting the brain (cranial trauma, meningitis, stroke, epilepsy, encephalitis), lesions of the central nervous system (ischemic events or haemorrhages), degenerative brain diseases (vascular dementia, Alzheimer's, Parkinson's disease, Huntington disease, Creutzfeldt-Jakob disease, neurosyphilis, Ménière's disease), metabolic imbalances

(Vitamin B12 deficiency, hypothyroidism, hypo- and hyperglycaemia, and low blood pressure), toxic substances (lead), alcohol and drug abuse, high fever, psychosis, panic attacks and depression. Source: http://www.my-personaltrainer.it/Sintomi/Disorientamento-temporale-e-spaziale. Accessed 6 September 2015.

10 On spatial disorientation in adults with brain damage, see also Vicari and Caselli (2010, Chapter 15, "Le competenze visuo-spaziali e visuo-costruttive").

11 www.faa.gov/pilots/safety/pilotsafetybrochures. Accessed 6 September 2015.

12 www.faa.gov/pilots/safety/pilotsafetybrochures. Accessed 6 September 2015.

13 The Orfield anechoic chamber was cited in the 2005 *Guinness Book of World Records* as the "World's Quietest Place".

14 The average distance travelled by children who have got lost between the ages of 1 and 6.5 years is 0.5 miles (0.8km).

15 The GDS is a seven-point rating scale used to assess whether a person has dementia-related cognitive impairments.

Part III

The labyrinth of the world

Places of disorientation

At the heart of this third section of the book is the powerful metaphor of the labyrinth, a metaphor that is frequently drawn on in art, philosophy, and literature to represent the confusion pervading the contemporary era and the difficulty of finding reference points in the "labyrinth of the world" we live in (Stiegler, 2009). The labyrinth has remained an enigmatic concept for centuries and still today acts as a powerful image of disorientation.[1] In the words of Jorge Luis Borges, considered one of the greatest contemporary labyrinthologists:[2] "There's no need to build a labyrinth when the entire universe is one" (Borges, 1970, p. 34).[3] Jacques Attali has argued that the labyrinth now belongs to the future:

> One of the keywords of tomorrow The microprocessor is already a labyrinth. Video games are also traversed by labyrinths, quests strewn with pitfalls. The same will apply to Internet, networks of power and influence, organigrams, degree courses, towns, and business careers. The digital fingerprint is a labyrinth that characterizes the identity of individuals. The brain is a labyrinth of neurons. Even psychoanalysis describes the unconscious as a monster concealed at the end of a labyrinth, and concerns itself with unravelling dreams in which the sleeper is faced with having to choose a path inside a maze of taboos. Learning to "think labyrinth" will become a necessity. Requirements: persistence, memory, intuition, curiosity. All qualities of the navigator or the nomad.
>
> (Attali, 1998, p. 193)

While in the first section of this book, I took the concept of space as my point of departure, and in the second centred my analysis around the geographical subject, in this third section I use places, understood as "meaningful locations" (Cresswell, 2014a, p. 12),[4] as a key to interpreting disorientation. Places bear a subjective value that can make them unique for the subjects who frequent them:

> But place is also a way of seeing, knowing and understanding the world. When we look at the world as a world of places we see different things. We see attachments and connections between people and place. We see worlds of meaning and experience. Sometimes this way of seeing can

> seem to be an act of resistance against a rationalization of the world, a way of seeing that has more space than place.
>
> (ibid., p. 18)

How can one orient oneself in the "labyrinth of the world", in which all is in movement, if place is: "an ever-shifting constellation of trajectories" (Massey, 2005, p. 151)? Viewing place as a key to interpreting disorientation prompts us to enquire in what places it is easiest to get lost and consequently, which places we might choose to best represent disorientation. In reality, as we have seen, disorientation is a subjective condition that can potentially concern any place; in certain circumstances, we can get lost anywhere, even inside our own homes (Alzheimer's disease). Thus, the next four chapters are partly the outcome of a subjective choice among many possible options, yet at the same time, they explore "exceptional places" that recur time and again in history, art, and literature, by virtue of their link with disorientation. First, in Chapter 7, I discuss the labyrinth itself, an ambiguous physical space in which order and disorder come face to face: "A labyrinth is a structure compounded to confuse men; its architecture, rich in symmetries is constructed to that end" (Borges, 1964, p. 110). In Chapter 8, enter the city, that place now inhabited by over half the world's population. The city may be the new form of labyrinth: "Another difference between the ancient and the modern labyrinth lies, as we have seen, in the fact that the ancient version is located in a place – castle, abbey, etcetera, – that is remote from the world, while the modern equivalent is right behind the curtains of this world, or even a disturbing metropolitan metaphor" (Orvieto, 2004, p. 87). Chapter 9 is devoted to the theme of wilderness, primarily in relation to the forest, which has served as a literary place of disorientation from Dante to Heidegger, but also in relation to the sea. Finally, in Chapter 10, I touch on the virtual labyrinth of cyberspace, "a hallucination consensually experienced on a daily basis", and on the enigmatic world of art.

Notes

1 The labyrinth, for Ingold: "is not available to the terrestrial traveler who is already embarked upon a journey across the earth's surface – a journey that is tantamount to life itself. The entrance to the maze marks the point not at which he touches down upon the surface, but at which he goes underground. Now as an interface between earth and air, the ground is a kind of surface that is visible from above but not from below. It does not have another side. Thus at the very moment of going underground, of entering the labyrinth, the surface itself disappears from sight. It appears to dissolve. […] the ghostly traveler finds himself in a world without any surface at all. Every path is now a thread rather than a trace" (Ingold, 2007, pp. 56–57).

2 *Enciclopedia Einaudi* (1979).

3 J. L. Borges, "Ibn-Hakim Al-Bokhari, Murdered in His Labyrinth", in *The Aleph* (1949); tr. Andrew Hurley, *Collected Fictions* (1998).

4 "Place acts as a category that serves to reproduce this normality and the existing 'order of things'" (Cresswell, 2014b, p. 251).

7 Labyrinths[1]

The figure of the labyrinth

The meaning of the labyrinth engages a broad area of enquiry that has produced a rich corpus of interdisciplinary literature spanning philosophy, art, archaeology, and mathematics (Matthews, 1922; Kern, 1982; Santarcangeli, 1984; Reed Doob, 1990; de Saint-Hilaire, 1992; Attali, 1996, Saward, 2002; Fisher, 2006). For Claude Raffestin: "The labyrinth provides us with a particularly rich metaphor for giving an account of the difficulties encountered in observing and explaining the world" (Raffestin, 1996, pp. 120–121). Still today, the labyrinth remains an enigma, while for centuries the meaning of its mysterious form has been the object of a variety of contrasting interpretations: Hermann Kern, in his detailed review of five thousand years of the labyrinth's history, emphasizes the ambiguity associated with the term in everyday language and points out its dual nature. While on the one hand, the labyrinth has featured since ancient times in both literary and metaphoric usage as a place in which one gets lost, up to the Renaissance it was graphically represented as a one-way path only. In Kern's view, the labyrinth can thus assume three different meanings: metaphor, maze, and labyrinth in the strict sense. Hocke asks in his study on Mannerism: "What essentially is the labyrinth in the ancient cultures? A 'unifying' metaphor for the predictable and unpredictable elements of the world" (Hocke, 1957, p. 101).

The metaphor of the labyrinth has been used since classical antiquity to describe situations of disorientation and confusion. Socrates declares in the *Euthydemus*: "but then we found ourselves in a sort of labyrinth, and when we thought we were at the end, when we had gone right through it, we came out again at the beginning, still as far as ever from a conclusion, just the same as when we first began to search" (Plato, *Euthydemus*, 291/b). As a metaphor, the labyrinth also sheds light on and enhances our understanding of philosophical and scientific theories. In the words of Raffestin: "The figure of the labyrinth has featured right throughout the history of Western thought; it is also a metaphor for aporia and this is why it is at the origin of Western thinking (aporia is a state of doubt due to the fact that there is more than one solution to a problem and all of the available solutions are equally valid)"

(Raffestin, undated, p. 2). In its metaphorical sense therefore, the labyrinth explores the passage from an "extreme condition of predisposition to doubt to a givenness with all doubt removed" (Blumenberg, 2009, p. 356), analogously to the method of Descartes. Bacon, in *Dedalus sive mechanicus* (1609) examines the ambiguous role of Daedalus and technology in the development of science: "The same person that built the Labyrinth made the Thread used by Ariadne to make her way out of the Labyrinth. The mechanical arts have ambiguous ends, simultaneously producing both the evil and its remedy" (Rossi, 2008, pp. 10–11). Blumenberg, in *Cave Exits* [*Höhlenausgänge* 1989], also explores the passage from the myth of the cave to the dream and the labyrinth. He comments that in Leibniz's *Nouveaux Essais*, the labyrinth is presented as the "imaginative background" against which "reality is the consequence of possibility", while Ariadne's thread is made to coincide with method:

> The thread is completely transformed into a technical instrument that reliably facilitates the analytical method of repeating unsuccessful attempts, when the nature of the problem prompts trying out all directions and coming back to the starting point each time, because there are no shortcuts connecting the different paths embarked upon.
>
> (Blumenberg, 2009, p. 364)

To get beyond the labyrinth, Leibniz was forced to rule out the concepts of absolute space and absolute time, becoming, according to Blumenberg: "… the first to develop the notion that the labyrinth was not the universe but reason, or that reason created the labyrinth 'around itself' only to find itself unable to make its way out" (ibid., p. 365). The multiple interpretations of "the impossible labyrinth" (Farinelli, 2003, pp. 198–199) lead Farinelli to conclude that:

> If we leave all rhetoric aside, the origin of the labyrinth is really very simple. It is the outcome of the collapse of the tower of Babel, the Babylonian ziggurat, the Cretan-Minoan palace, the Egyptian pyramid: in short, the collapse or razing to the ground of all vertical structures whose various models correspond to a hierarchical system of power. In other words, it is the product of the transformation of the world into Earth. Thus, the different levels mutate into horizontal dimensions placed recursively one inside the other, and it is this recursivity that prevents us from speaking of space.
>
> (ibid., p. 21[2])

In the notion of the labyrinth, this author contends, our spatial reference points and coordinates are challenged and overturned: implicit in it therefore is the fundamental contradiction of cartography. What is a labyrinth? Farinelli reflects that:

> The labyrinth is the nightmare of Western culture, because it is impossible to map. If you try to map a labyrinth, you transform it into quite another thing, into its contrary: something which has a centre, while the centre is exactly what does not exist in a labyrinth. This happens because the labyrinth is not a thing but a condition, based on the impossibility to image the centre, and on the necessity to find it. In other words, what is at stake when we draw or build a labyrinth is exactly the mystery of objectification, the transformation of a situation into a thing. According to modern Western culture, knowledge is representation, i.e. possibility to map, which means that it is impossible to know what is impossible to represent. This is why the labyrinth is for us the symbol of the negation of knowledge, a sort of brain cramp. However, this is so only because we have forgotten the true nature of the labyrinthine model.
>
> (Farinelli, 2002, pp. 79–80)

Kern proposes that in addition to being a metaphor, the labyrinth combines two ideas that are in contrast with one another in terms of both form and content – the "monocursal" labyrinth and the "pluricursal" labyrinth or maze – leading to an obvious confusion of meanings that nonetheless only began to clearly emerge at the end of the twentieth century (Kern, 1982, p. 13). The first type, that is to say, the labyrinth in the strict sense, is for Kern, a strongly visual concept – invariably monocursal – lending itself to linear graphic representation as a geometric shape, either round or quadrangular. Viewed from above, the diagram of the labyrinth looks like a path, delimited by curved or angular walls that, starting from a single side opening (usually facing West), mark a fixed path leading to the centre; the centre-point does not have a way out and is the starting-point from which to return towards the outside. This form, common to many cultures, decorated Cretan coins during the Minoan era. The second type of labyrinth, the maze, is an intricate structure with false exits, which may be either a garden or a building; the visitor is continuously confronted with choices regarding the way to go, running the risk of coming to a dead end and getting lost. This type of pluricursal labyrinth, which has existed since antiquity in both literature and metaphor, only appeared in figurative art in the mid-fifteenth century.

The ambiguity of the concept is particularly explicit in Italian and the other Neo-Latin languages, which do not distinguish between a labyrinth in the sense of monocursal structure on the one hand, and an entanglement of paths, or pluricursal or Mannerist labyrinth (*maze*) on the other. Thus in the Romance languages, a single term, labyrinth, is used for both meanings, while in the Germanic world there are usually two different terms reflecting the distinction between monocursal and pluricursal labyrinths. In English, the term *labyrinth* exists alongside the Anglo-Saxon word *maze* which "comes from the Welsh *maes* and means 'pasture' in Cornish; or from the Celtic *maes*, 'field'. [...] In English, the expressions *amaze* and *amazement* indicate a state of surprise and confusion such that the person experiencing it does not know which way to turn" (Santarcangeli, 1984, p. 46) (see also the relationship with the word *amuse*, discussed in the Introduction §2). In German, in addition to *Labyrinth* we find *Irrweg* and *Irrgarten*, both related to the verb *irren*, with the multiple meanings of "go wrong", "make a mistake", "wander" and "go mad". The first visual representation of an *Irrgang* appears in the parchment collection of the Paduan doctor Giovanni Fontana, dated to around 1420 (Kern, 1982).

Paolo Santarcangeli has defined the labyrinth as "a tortuous path, along which it is sometimes easy to lose one's way without a guide", also distinguishing between different types: "a) an edifice that is impossible to get out of without guidance; b) a cave housing an intricate maze of tunnels; c) an ingeniously designed path that is scattered with elements of deception, or involves a very complex route; d) a metaphor" (Santarcangeli, 1984, p. 173). Kern, in another attempt to bring order into this broad and complex subject matter, identified three different contexts in which the labyrinth may be found: the labyrinth as a literary motif, in this case appearing under the form of a system of false paths, or *Irrgang*; then, the labyrinth as a figure of movement or dance; finally, the labyrinth as a graphic figure in drawing, mosaics, and engravings (Kern, 1982, p. 22). In Kern's view, the labyrinth is not to be confused with other similar graphic representations, such as the spiral, meanders, or knots in the drawings of Leonardo da Vinci or Albrecht Dürer, themselves the object of complex symbolic interpretations in the spheres of art and literature. The relationship between the labyrinth and the mandala is also highly complex, given that the latter is sometimes represented with a labyrinthine structure and in some regions is used in dance (Fanelli, 1997). Finally, Umberto Eco suggests the organization of the labyrinthine form into three distinct categories, following Deleuze and Guattari by adding a third dimension to the classification scheme of unicursal and Mannerist labyrinths: the rhizome. The labyrinth in the strict sense, or unicursal labyrinth, is presented by Eco in the following terms:

> In the first place, there are different types of labyrinth [...] The first of these is the so-called "unicursal" labyrinth: when viewed from above it seems indescribably intricate and when going through it one is overcome by the fear of never getting out [...] This is the classical labyrinth, which actually does not need Ariadne's thread because it is its own Ariadne's thread. That is why the Minotaur has to be at the centre of it, to make the whole affair less boring. The problem posed by this labyrinth is not "How will I get out?" but "Will I get out alive?" This labyrinth is the image of a cosmos that is difficult to live in, *but* on the whole ordered (it has been conceived by a mind).
>
> (Eco, 1984, p. XIII, original italics)

Dance, dance, otherwise we are lost

As an esoteric symbol, metaphor for the world and complex construction, the labyrinth points on the one hand to loss of meaning, the absence of limits and the abandonment of all known reference points, and on the other to the building of a perfect, complex, and detailed control mechanism (designed to keep monsters and, more generally, irrational fears in check) and to transition (in the sense of a rite of initiation). Labyrinths, according to Reed Doob, author of *The Idea of the Labyrinth*, are the place of "…enforced circuitousness; disorientation; the idea of planned chaos; the *bivium* or critical choice between two paths, inextricability; intricacy; complexity" (Reed Doob, 1990, p. 2); the ambiguity of the labyrinth is expressed by the dual perspective of those treading it, who are confused by being unable to see ahead or behind of them, and those viewing it from above, who are "dazzled by its complex artistry" (ibid., p. 1). The two categories simultaneously "incorporate order and disorder, clarity and confusion, unity and multiplicity, artistry and chaos" (ibid., p. 1). The same ambiguity is to be found in dance, which, on the one hand provides an unlimited space of disorientation thanks to its subversive and liberating potential – from electronic music and rave parties to the states of trance induced by tribal and religious (Sufi) dance – but on the other, offers perfect choreography, harmonic form, style, discipline, and preparation. Thus, a key parallel may be drawn between the labyrinth and dance.

The space of dance and the richness of its implications regarding the centrality of the body and the language of performance are well-established areas of enquiry within the new cultural geography: "Along with the new language of performativity the vocabulary of dance is being enlisted in order to rethink ideas of subjectivity, embodiment and social identities" (Nash, 2000, p. 653). The theme of dance also speaks to the discipline's growing interest in the body as a geographical subject, the "complexity and uncertainty of places and performed spaces" (Gregson and Rose, 2000, p. 434) and the new prospects opened up by non-representational theory as an instrument for spatial analysis (Davis and Dwyer, 2007; Nash, 2000; Delgado and Muñoz, 1997; Law, 1997; Thrift 1997, 2008). For geographers: "Dance presents representations of bodies in spaces, their relations to the spaces and to other bodies, and […] in this sense it is a most pertinent arena for exploring questions of subjectivity. […] The similarities and differences of dancing bodies point to the kinds of subjects performing. They are seen as gendered, 'racialized', sexualized and in terms of their ability" (Briginshaw, 2009, pp. 5–6). Numerous case studies have explored the relationships between space and dance. In analyzing the growth in popularity of *salsa* in North Carolina, Johnson explains that: "Dance, as a movement and performance of and between bodies, creates and is created by spaces that are fluid, multiplicitous, and political" (Johnson, 2009, p. 4). As well as body and movement, dance does indeed bring into play political and cultural values regarding social, ethnic, gender, sex, and ability differences in the bodies performing it: "Much of the work on dance within Geography has focused on connecting dance as cultural expression to identity, nation, citizenship, and public space […] Scholars have also argued that dance has the ability to defy, evade, and subvert oppressive power structures and categorizations" (Johnson, 2009, p. 101).

According to Nigel Thrift (1997), dance and its potential as an alternative practice of resistance occupy a key role within non-representational theory (NRT). Thrift's interpretation of dance as a "metaphor for unreflected, unarticulated, practical action" (Revill, 2004, p. 201) is comparable to the notion of the labyrinth as a place of chaos. For Caroline Nash: "At least some of the appeal of dance in day dreams or theory is the idea of escape from (self)consciousness and the limits and demands of the social world, to lose 'oneself' in the motion, to find a realm unmediated by thought and the burdens of consciousness and individualism – to dance yourself dizzy" (Nash, 2000, p. 657). This experience of abandon and uncontrolled movement may be found in contemporary disco-dancing, the insistent beat of electronic music (techno, acid house, jungle), rave parties, tribal dances, and psychedelic or religious trances. In relation to the meditation of the whirling dervishes, Njaradi comments that:

> We can see that contemporary dance elements are referred to purely as a "technique" while whirling is connected to religion and purification. […] Sufi dance elements are perceived as inherently "out of order", "ecstatic", "messy", whereas contemporary dance elements are there to give it a "necessary structure" in order to make it an "art work" rather than religious ceremony.
>
> (Njaradi, 2008, p. 25)

However, the interpretation of dance as total abandonment to the rhythm of the music has been critiqued (Revill, 2004; Nash, 2000; Thein, 2005), because it fails to take account of the political, social, and cultural leanings that may be expressed through dance, as well as the formalism and dedication required to execute it. Dance can have a liberating and transgressive power but it can also demand discipline, order, and rigorous limits (Revill, 2004; Cresswell, 2006; Markula and Pringle, 2006; McMains, 2006; Gottschild, 2003). Revill, in a field study on French folk dancing, emphasized the years of training required and his own failure as a novice to master the precise geometry of the steps and movements. In a description of his initial attempts to dance the *bourrée*, he wrote:

> As a beginner, I found it almost impossible. […] I gripped my partner too tightly because I was concentrating too hard on my feet, I moved boldly forwards when everyone else moved back. I began to realize that even the most basic movements only become "natural" by habituation. In addition to this, I had severely underestimated the extent to which dancers need to develop a memory able to map out the movement of the body in space and time. Dancers need to sense aurally and rhythmically where they are

> in the music at the same time that they have sufficient spatial awareness to know where they are on the floor, where they are going.
>
> (Revill, 2004, p. 204)

Likewise, Saldanha, in an analysis of rave parties on Anjuna Beach in Goa, observes that Thrift's approach addresses the "directness of the sensory experience", but "gives scant attention to the differentiations between bodies, and therefore to the actual politics that arise in spaces like those of dance" (Saldanha, 2005, p. 717), thus missing out on the rich social implications that emerge from music. The construction of subjectivity arises within the relationship between space and dance:

> These key interrelated ideas, current in critical theory, are the blurring of bodily and spatial boundaries, the in-between spaces of hybridity and ambiguity, a focus on inside/outside interfaces, privileging touch and sensation over the visual, focusing on foldings of bodies and space, and on bodily and spatial excesses. When dance is examined using such ideas, the mutual construction of bodies and spaces is at the centre of the investigation.
>
> (Briginshaw, 2009, p. 20)

The rich symbology of the labyrinth may be used to address issues of subjectivity and disorientation in the contemporary era: "The labyrinth is also part of the modern world, and it is in this context that the labyrinth raises questions of actual practice – of movement and identity, gender and fear" (Salazar, 2003, p. 6). With regard to dance, the themes of identity, subjectivity, and the body are not only highlighted by geographical discourse, but also feature strongly in contemporary choreographic practices. An example of this is the work of the great choreographer and dancer Pina Bausch and her dance company, the Tanztheater Wuppertal:

> The impact of her choreographic imaginary is enhanced in many of her dances by the particular kinds of resonant space she creates on stage. [...] Performers are often seen struggling with the elements – snow, mud, wind, water. Sometimes they appear distressed or lost, crying out or shouting to attract attention, wandering aimlessly through woods, down roads or over hillsides.
>
> (Briginshaw, 2009, p. 8)

Labyrinthine wandering is a recurrent theme in the work of Pina Bausch. In Wim Wenders' 2011 film *Pina*, the dancer Silvia Farias Heredia of the Wuppertal Dance Theatre comments in an interview: "Sometimes she would say things like: 'Go and search!' But that was all she would say. It meant that you were to keep on searching without knowing where to look, or whether you were on the right track. Pina was a radical explorer. She would look deep into our souls. There was one theme in particular, she kept asking us: 'What are we searching for? Where does all this yearning come from? Dance, dance, otherwise we are lost'."

The dance of the labyrinth

The labyrinth, from its very origins, has been connected with dance. Carl Kerényi wrote that: "All research on the labyrinth ought to begin with dance" (Kerényi, 1983, p. 56). Paolo Santarcangeli reached the same conclusion: "Thus, there is a constant association between labyrinth, cave and dance" (Santarcangeli, 1984, p. 158). The mysterious figure of the labyrinth-dance appears on a number of archaeological remains, the most ancient of which is an Etruscan terracotta wine-jar (*oinochoe*) from Tragliatella, Italy (620 BC) depicting dancing figures and horsemen. Dance is also closely linked to the myth of the Cretan labyrinth: Daedalus, architect and inventor at the court of Minos in Knossus, made Ariadne a gift of a *choros* – a dance floor or perhaps, according to an alternative interpretation, the steps of a dance. Theseus, having killed the Minotaur and made his way out of the labyrinth, sailed away from Crete, landing on Delos, where he and the youths he had rescued performed the *geranos*, or crane dance. Theseus' dance later featured in the works of a number of classical authors. Homer, in the *Iliad*, describes it as represented on the shield forged for Achilles by Hephaestus. Virgil, in the Aeneid describes the horse-back riding display carried out by the Trojan youths for the funeral rites in honour of Anchises: the labyrinthine *lusus Trojanus*.

The labyrinth-dance appears to have given rise to the Candiotta, which is still danced in modern-day Greece, as well as to the Scottish folk dance known as the maze, and to labyrinthine dances found in the Middle East, Romania, and Sardinia. The use of the labyrinth in dance and in equestrian or other types of display, is an ancient European tradition alluded to by a vast assortment of authors, from Rostand to Shakespeare ("Here's a maze trod indeed. Through forthrights and meanders!" *The Tempest*) or Milton. Dancing took place in the Northern European *Troiaburg* and on the floors of cathedrals. As Robert Graves explained in his essay on the myth *The White Goddess*:

> In Great Britain the tradition of the Spiral Castle survives in the Easter Maze dance of country villages: the mazes being called Troy Town in England and in Wales Caerdroia. The Romans probably named them after the Troy Game, a labyrinthine dance of Asia Minor, performed by young noblemen at Rome under the Early Empire in memory of their Trojan origin […] The maze dance seems to have come to Britain from the Eastern Mediterranean with the New Stone Age invaders of the third millennium B.C.
>
> (Graves, 1948, p. 127)

Thus, movement is always implicit in the labyrinth; its trajectory does not express the static condition of a physical place but an initiatory path. Giorgio Ieranò has described it as a: "rite of passage from the profane to the sacred, from the ephemeral and illusory to reality and eternity" (Ieranò, 2007, p. 28). Greene, in an examination of Renaissance documents and choreographies of dances or equestrian games at the French Valois Court, proposed metaphysical interpretations of the labyrinth, associated with magic, Neoplatonism, and orphic cosmogony (Greene, 2001). In the Middle Ages, it was customary during the Easter festivities to dance in labyrinths – invariably unicursal – built into the flooring of the French cathedrals, perhaps as a sign of purification in accordance with the teachings of medieval scholasticism: "The world is a labyrinth in which the soul must wander until it finds liberation" (Hippolytus, 3rd century AD). Making one's way through a labyrinth involves regular rhythmic movement, with a single entrance, a single exit, a centre to start back from and the need to return to the starting point. The interpretations of this type of movement are many and various, and include initiation, death, the descent into hell, and rebirth. Starting from the hypothesis that *Labyrintos* was originally the name of a dance, and only later fixed in the graphic figure, Kern revisited the key question about the labyrinth – is it form or chaos? – asking how regular movement and a rhythmic cadence of steps (unicursal labyrinth) may be transformed into unseemly agitation (Mannerist labyrinth); however, his analysis failed to consider the transgressive and inebriating power of music.

The Cretan labyrinth

The labyrinth is believed to have originated on Minoan Crete during the Neolithic Age, from 3000–2000 BC onwards. The Glastonbury Tor labyrinth in England dates from the same period, lending support to the nineteenth-century theory that the labyrinth originally came from Northern Europe (Krause, 1893, Matthews, 1922), a theory that Santarcangeli also argued "may not be rejected with certainty at the current state of knowledge" (Santarcangeli, 1984, p. 93). In any case, the key question of how the labyrinth came to feature in different cultural contexts remains to be fully explored. For Kern, the basic labyrinthine form, a continuous circular path containing seven spirals, is only found in specific and well-defined cultural contexts: Europe, India, Java and Sumatra, and the south-western United States. Whether the archetype of the labyrinth corresponds to a fundamental human need, an *Elementargedanke*, as hypothesized by Adolf Bastian (1895), or has a single shared origin in the labyrinth of Minoan Crete, which may have subsequently emigrated elsewhere in line with Friedrich Ratzel's *Migrationtheorie* (1882), is still open to debate. The etymology of the word labyrinth is also debated and mysterious: according to Kern, the earliest reference to the labyrinth may be a Cretan inscription in Linear B, dated 1.400 BC and found in Knossos, that calls for the offering of: "An amphora of honey to all the gods, to the Lady of the Labyrinth, an amphora of honey." The German archaeologist Mayer was the first in 1892 to suggest a link between the term labyrinth and *labrys*, or double-headed axe, a Dionysiac symbol. This etymology was also mooted by Arthur Evans on the basis of his excavations at the palace of Knossos in which the double-bladed axe was a recurrent feature; thus, the labyrinth could be the house of the double axe. However, according to Kern and Kerényi, this theory no longer stands up since the deciphering of Linear B clarified that the term *peleky*, and not *labrys* was used to refer to the double-headed axe in Minoan Crete.

Other scholars have linked the term to the underground world: "It may be related to an Indo-European root to be found in words such as the Greek *laas* or the Latin *lapis* ('rock', 'stone'). If this were so, labyrinth would mean 'house of stone' or 'house in the rock'" (Ieranò, 2007, pp. 23–24); along these lines, many ancient authors identified the intricate cave system near Gortyn, at the foot of Mount Ida, as the site of the Cretan labyrinth and the Minotaur's underground prison. Strabo described the caves at Nafplio in Peloponnese as "Cyclopic labyrinths"; this could imply a link between cave and labyrinth, and specifically that labyrinths might have a religious origin linked to cave sanctuaries. Consequently, the labyrinth is related to Hades: "The labyrinth in short represents the bowels of the earth, the image of the Hereafter translated into the enigmatic cipher of meanders and spirals" (Ieranò, 2007, p. 27) For Kerényi too: "Labyrinthine and underworld are one and the same thing, a single origin" (Kerényi, 1983, p. 34), whereas Kern and Santarcageli have taken a less decided stance on this connection. Mircea Eliade, after emphasizing that the labyrinth is such a complex concept that it cannot be distilled into a single interpretation, went on to observe that caves have fulfilled a religious function since the Paleolithic Age:

> In prehistoric times the cave, often assimilated to, or ritually transformed into, a labyrinth, was both the theatre in which initiations took place and the place where the dead were buried. In its turn, the labyrinth was equated with the body of Mother Earth. Going into a labyrinth – or a cave – was the equivalent of a mystic return to the Mother.
>
> (Eliade, 1976, p. 145)

Eliade also linked worship of the mother-goddess with the city of Troy, a name often connected with Northern European labyrinths: "Troy was considered on a par with the earth, that is to say, as a goddess; the inviolability of the ancient cities was likened to the virginity of their goddess protectress" (ibid.). The name Troy, which frequently crops up in reference to British and Scandinavian labyrinths (Troy town, Caerdroia, Troja, Trelleborg, Trö, Trojeborg, Tarha), may be related to the Old German term *drajan* or the Celtic *troo* "turn " (Matthews, 1922). The writings of the ancient classical authors – Strabo, Apollodorus, Philostratus, Diodorus, Plutarch, Callimachus, Hesychius, Ovid, or Pliny – transmit great uncertainty regarding what exactly is signified by labyrinth, whether cave or symbol or construction in which one loses one's bearings, offering a plurality of meanings that only serve to heighten the confusion. In classical culture, the labyrinth was linked above all to the Cretan myth and to the entangled relationships among the protagonists. As Raffestin has explained:

> The labyrinth is a complex figure used by historians and geographers to indicate complicated buildings of uncertain purpose. In Latin and Greek literature, the motifs that may be pointed up by this figure are related to the myth of Theseus, Ariane, Daedalus and the Minotaur. In the Greek language, Herodotus is the first to use the term to refer to the Egyptian building with three thousand rooms whose construction was ordered by Pharaoh *Amenemhet* III to the south of his pyramid in Hawara, near Lake Moeris. Strabo and Pliny the Elder further claim that Daedalus drew inspiration from this structure when building the Cretan labyrinth.
>
> (Raffestin, undated, p. 2)

The Egyptian labyrinth, cited by Herodotus, Strabo, Pomponius Mela, and Pliny the Elder, was an extraordinary construction, of admirable size and complexity, and full of dead ends that were guaranteed to make one lose one's bearings. The classical authors also emphasized an Egyptian influence on the work of Daedalus, creator of the Cretan labyrinth. This ties in with Santarcangeli's comment that: "If not entirely certain, it is more than probable that during the ancient Kingdom there were close cultural ties between Egypt and Crete, for several reasons" (Santarcangeli, 1984, p. 50). In Kern's view on the other hand, the Hawara labyrinth, located near Krokodilopolis, was predominantly intended to be a funeral monument. Herodotus described it, having actually visited it around 450 BC, as a building of 3,000 rooms on two levels, constructed around 12 courtyards and surrounded by walls: "The ways leading through the rooms and the twisting and turning paths through the courtyards, which are highly intricate, inspired great awe in those who passed from the courtyard into rooms, and from the rooms to porticos and from the porticos into other rooms and from those rooms into other courtyards" (*Storie*, II, 149). For Strabo, who lived during the first century BC, this was the first example of a labyrinth, built around 1,600 years before his time: "Before the entrances to the halls there are a great number of long and tortuous galleries that are intertwined with one another in such a way as to prevent a stranger from entering or leaving without a guide" (*Geografia*, XVII, 1, 37, 5 and 12). Again, Pliny the Elder in his *Naturalis Historia* (first century AD) cited the Egyptian labyrinth as a model for all labyrinths, a model that is clearly of the Mannerist variety:

> A circular entanglement of passages that wind, advance and retreat in a bewilderingly intricate manner. It is not just a narrow strip of ground comprising many miles of walking such as we see in our mosaic floors or

> in the cross-country game played by our boys, but a real building with doors let into the walls at frequent intervals to deceive with regard to the way ahead, and force the visitor to retrace his steps [...] It is not possible to describe the plan and its different areas.
>
> (Pliny, pp. XXXVI, 84–86)

The story of the labyrinth is structured around the affairs of the protagonists in the Cretan myth. In Hans Blumenberg's view, the stories in the myth: "were not told to answer questions, but to dispel hardship and dissatisfaction, which constitute the first condition necessary for questions to arise. If fear and uncertainty are to be addressed, either they must be cut off at source or the questions about what is causing and fuelling them must be made explicit" (Blumenberg, 1991, p. 231). The key to the dark meaning (or at least to one of the possible meanings) of the complex labyrinthine structure and the deep existential drives behind its invention is to be found in the intricate relationships among Daedalus, Minos, Theseus, Ariadne, and the Minotaur. The architect of the labyrinth, Daedalus, makes a triple contribution to the myth: first he built the apparatus that allowed Pasiphae to mate with the bull; then, his architectural skill and experience came into play in designing the labyrinthine structure, a sophisticated work of engineering designed to hide the impious result of Pasiphae's infidelity: "Whether it is the 'house of the double-edged axe' or 'the house of stone, palace or cave', the labyrinth of Knossos is invariably identified as a prison for the monstrous man-bull, the hybrid Minotaur" (Ieranò, 2007, p. 31). Its purpose is to disorient all who enter it, preventing them from ever finding a way out. Third, Daedalus gave Ariadne the *choros*, a rhythmic pattern of steps or, perhaps, a place in which to carry out such a ritual dance. Thus Daedalus' labyrinth contains both key conceptions of the labyrinth explored by Reed Dobb, which as we have seen, are also to be found in dance: on the one hand, an extraordinary and sophisticated construction project, and therefore the height of formalism, and on the other, total loss of orientation and the headiness of abandoning one's reference points.

From the medieval to the Mannerist labyrinth

The motif of the labyrinth, in a variety of shapes and sizes, featured in many medieval churches, in North Africa, Italy, and especially in France. This type of labyrinth is not found in England, which on the contrary boasts numerous *turf mazes*, or country labyrinths, cut into lawns or stone circles, and common to all of Northern Europe. The popularity of church floor labyrinths reached its peak in the twelfth century, although the oldest known instance, located in the Basilica of Saint Reparatus in Chlef in Algeria, dates from 324 AD. Many of these labyrinths have been lost; the most noteworthy of those that have been preserved are in the cathedrals of Amiens, Bayeux, Lucca, and Chartres. The latter is perhaps the most famous: it consists of a unicursal path with 11 circular spirals, measures almost 13 meters in diameter and is related to the lunar calendar. In other churches (Reims, Amiens), the labyrinths are octagonal in form. The floor structure was called "maze", "meander" or "Via Dolorosa", while its centre was termed "paradise" or "Heavenly Jerusalem" (Wright, 2001). A shared account of the meaning of the labyrinth is not to be found in medieval church writings. Nor does the labyrinth feature in proto-Christian architecture or in the catacombs. It may be interpreted as a symbol of the difficulty of walking towards Christ or the finite nature of the world, as a symbolic substitute for the pilgrimage to the Holy Land, a symbol of masonry or of initiation into a confraternity (Matthews, 1922). The theme of movement recurs yet again in the medieval churches: the floor labyrinths were variously used for dancing, ball games, and ritual ceremonies, especially at Easter time (Santarcangeli, 1984, Hayes, 2003). Raffestin has further commented that:

> In the Middle Ages, the labyrinth is the world of evil and the most complete form of this world is hell. While the medieval labyrinth is sinister, during the Renaissance, the labyrinth becomes tension between the external and the internal: it moves from objective to subjective; external space corresponds to internal space as does macrocosm to microcosm. In the seventeenth century, Jan Amos Komeńsky used the figure of the labyrinth to express the impossibility of understanding the social world.
>
> (Raffestin, undated, p. 2)

Komeńsky or Comenius referred, in his book *Labyrinth of the World and Paradise of the Heart* (1631), to the model of the Cretan labyrinth and its rich symbology. Deeply disappointed by the policies of the Counter-Reformation in Bohemia, in his satirical allegory Comenius had his Pilgrim wander about the labyrinthine city before finally finding himself through his new relationship with God. The guide Ubiquitous asks the Pilgrim: "Hast thou heard of the Cretan labyrinth?" [...] "It was a wonder of the world, a building consisting of so many chambers, closets and corridors, that he who entered it without a guide walked and blundered in any direction, and never found the way out. But this was nothing compared to the way in which the labyrinth of this world is fashioned, particularly in these times" (Komensky, 1901, pp. 61–62). Thus, the city described by Comenius was in contrast with the model of the ideal city and was presented "in the guise of labyrinth as the 'opposite' to utopia, insofar as in the *Labyrinth* everything is 'wrong'" (Stroppa, 2001, p. 24). Here, the metaphor of the labyrinth once more takes the form of a binary opposition, between good and evil, order and chaos.

Over time, the symbolic-religious meaning of the medieval labyrinth gradually weakened as the figure took on a more secular and playful meaning, while the construction became more complex, featuring more and more blind alleys. Many references to the labyrinth in the medieval and Renaissance literature (Petrarca, Boccaccio, Ariosto, Tasso) attribute it with an allegorical meaning, as a "metaphor of love that inhibits and destroys" (Orvieto, 2004, p. 35) and particularly in relation to sex as "labyrinthine mental confusion" (ibid., p. 40): "The labyrinth represents the place of confusion, of the enigma in which one is trapped either through the work of sorcerers and demons, or by one's one hand because subject to the passions of the flesh [...] the labyrinth of love, or of error in general, is predominantly, at least in literary contexts, the *multicursal* kind ('with several paths')" (ibid., p. 35). Hence the labyrinth no longer offered the rhythmic cadence of a unicursal path but had become decidedly Mannerist, a maze of choices, false paths, error, and disorientation. In the words of Umberto Eco: "The second type of labyrinth is the Mannerist kind: if you unwind the classical unicursal labyrinth you will be left with a thread, but if you are able to disentangle a Mannerist labyrinth you will not end up with a thread but with a tree structure, with infinite ramifications, ninety-nine percent of which lead to a dead end" (Eco, 1984, p. XIII).

The labyrinth became an intellectual game and an entertainment, though without shedding its mystic, magical, and allegorical baggage: "From around the mid-sixteenth century [...] representations of the labyrinth became increasingly more common, attaining in the sixteenth-seventeenth centuries a wealth of iconography without precedent, and at the height of the Baroque period, a degree of ubiquity that was bordering on obsession" (Santarcangeli, 1984, p. 213). In the sixteenth and seventeenth centuries, the labyrinth began to appear in the villas, castles, and palaces of Europe as a *jardin de plaisance*, a sophisticated composition of hedges, flowerbeds, bushes, and paths. The best-known examples of labyrinths include: the Villa d'Este gardens in Tivoli, designed by Pietro Ligorio in the mid-sixteenth century and commissioned by Hyppolytus d'Este; the labyrinth of Versailles, designed by Le Nôtre from about 1662, with allegorical fountains and sculptures inspired by Aesop's fables, and later destroyed in 1775; and, above all, the Hampton Court maze in Surrey,[3] created by George London in 1690 for William III of Orange.

The labyrinth is cited in Italian, French, and English gardening manuals as a highly desirable ornament (see, for example Thomas Hill alias Didymus Mountain [1577], *The Gardener's Labyrinth*). The fashion of labyrinths in private gardens declined at the end of the eighteenth century and the labyrinth became a popular feature in public spaces as a form of amusement. The labyrinth game also began to be introduced into closed spaces, in the context of exhibitions, funfairs, and amusement parks; the oldest known labyrinth of mirrors, created to celebrate the 1891 Jubilee, is still on view today at the fairgrounds on Petrin Hill in Prague (Fisher, 2006). In the contemporary era, the labyrinth has not lost any of its charm, continuing to be offered as an attraction, in a wide variety of forms, all over the world. We live in the age of *edutainment*, in which education meets leisure time: "Thus mazes and labyrinths are appearing in shopping malls, city centers, waterfronts, museums, safari parks, zoos and aquariums, churches and schools" (ibid., p. 18). At Warren Street[4] underground station in London, a ceramic mural labyrinth keeps passengers entertained during their three-minute wait for the next passing train, which coincides with the average length of time required to solve the maze.

Hedge labyrinths are costly to maintain and so there is a recent tendency to substitute them with labyrinths built of wood panels. The model created by Stuart Landsborough, on a small island off New Zealand, set off the 1980s *maze craze* in Japan: within a few years over 200 wooden labyrinths had been built, attracting visitors from all over the country. In the US, outsize labyrinths have been entered in the Guinness Book of Records.[5] The idea of cutting labyrinths into cornfields is credited to the Broadway producer Don Frantz of Pennsylvania and to Adrian Fisher: the first "maze of maize" in 1993 was an enormous dinosaur covering an area of almost 1.5 hectares, with a path almost 3 km long; corn mazes have been extremely popular, and have become progressively bigger, attracting thousands of visitors during the annual season (ibid.). The biggest labyrinth in the world today – art publisher Franco Maria Ricci's *Labirinto della Masone*, inaugurated in May 2015 – is located in Italy. Ricci, a friend of Italo Calvino, Roland Barthes, and Jorge Luis Borges, has a longstanding fascination with the ambiguity of the labyrinth. The star-shaped Labirinto della Masone covers an area of seven hectares and is entirely constructed of different species of bamboo plant. There is a pyramid at the exit, which symbolizes the Catholic Trinity but also freemasonry, revolutionaries, secularism and mystery in general. Finally, in the virtual world, the increasingly sophisticated simulations available for computers and videogames open up new prospects for and add new dimensions to the idea of labyrinth. Hence, the labyrinth, in all its forms, remains a powerful metaphor for interpreting the contemporary world. Let us thus leave the last word to Jacques Attali: "Understanding the labyrinth will soon become crucial to mastering modernity. [...] From Daedalus to the Internet ... Very far? And very near. Just like two neighbouring points in a labyrinth" (Attali, 1996, pp. 22–23).

Notes

1 Part of this chapter, here revised and extended, was previously published in Schmidt di Friedberg (2016).

2 The labyrinth is a key theme in the work of Franco Farinelli. See also Lladò (2013).

3 An amusing account of getting lost in Hampton Court Maze is provided by Jerome K. Jerome (1889) in *Three Men in a Boat*. A model of the Hampton Court Maze was used to study orientation abilities in animals, in experiments conducted with rats, first by experimental psychologist William Small in 1899, and later by Edward Tollman (see § 2.3).

4 Warren Street is named after Anne Warren (1737–1807), daughter of Admiral Sir Peter Warren. However, in English, the word "warren" means a maze of tunnels dug by rabbits, a place in which it is easy to get lost.

5 The world's largest temporary corn maze measured 163,853.83 m^2 (40.489 acres) when opened to the public in September 2007 at Coolpatch Pumpkins in Dixon, California, USA.

8 The city

A labyrinth where you are never lost[1]

The broken map

"The city – a bounded infinity. A labyrinth where you are never lost. Your private map [...]. Even if you lose your way, you cannot go wrong" (Abe, 2001). This quotation from a novel by Kobo Abe provides the point of departure for our investigation of the city as a place of disorientation. Kobo Abe's book, *The Ruined Map* (*Moyetsukita Chizu*, 1967) leads us into the intricate windings of a boundless megapolis, through a maze of streets and new neighbourhoods: the lead character in this novel is, in no uncertain terms, 1960s Tokyo: hyperactive, violent, and engaged in a tumultuous and disorderly process of expansion. It is the city "without addresses" explored by Roland Barthes in his *Empire of Signs*, a city whose very map is undecipherable:[2]

> The streets of this city have no names. There is of course a written address, but it has only a postal value, it refers to a plan (by districts and by blocks, in no way geometric); knowledge of which is accessible to the postman, not to the visitor: the largest city in the world is practically unclassified, the spaces which compose it in detail are unnamed. This domiciliary obliteration seems inconvenient to those (like us) who have been used to asserting that the most practical is always the most rational (a principle by virtue of which the best urban toponymy would be that of numbered streets, as in the United States or in Kyoto, a Chinese city). Tokyo meanwhile reminds us that the rational is merely one system among others.
>
> (Barthes, 1982, p. 33)

In Abe's novel, a private investigator is engaged by a mysterious woman to find her husband, a man called Nemuro, who has been missing for over six months. Armed with only two clues – a photograph and a matchbox – the nameless detective embarks on a search that leads him through the criminal underworld of Tokyo. He wanders about the streets of a metropolis that he increasingly perceives as unfamiliar and unreal, eventually coming to identify so strongly with the missing person that he can no longer discern the boundaries of his own identity. Similarly, in Paul Auster's *City of Glass*, the detective's identity coincides with the map of the city and its organized system of well-known places, a map that gradually becomes fragmented to the point of disappearing completely. New York, in the words of the novelist: "was an inexhaustible space, a labyrinth of endless steps, and no matter how far he walked, no matter how well he came to know its neighbourhoods and streets, it always left him with the feeling of being lost. Lost, not only in the city, but within himself as well" (Auster, 1985, p. 8). Like Auster's character Quinn, who gets lost in a labyrinth that is located both in space and in his mind, Abe's detective wanders through the seedy areas and new districts of a city that no longer offers reliable landmarks, ill-assorted pieces of a jigsaw puzzle that cannot be put back together: Nemuro is never found, and the investigator, in turn, loses himself, ultimately becoming unable to recognize even familiar and well-known landmarks: "I too left my crevice in the darkness and began walking in the opposite direction [...] relying on a map I did not comprehend" (Abe, 2001, p. 299).

Tokyo – in terms of population, the largest metropolitan area in the world today, with 38,140,000 inhabitants (UN, The World's Cities, 2016)[3] – introduces us to the theme of the urban labyrinth. As Barthes describes it, the Japanese capital does have a centre but this centre is empty:

> The entire city turns around a site both forbidden and indifferent, a residence concealed beneath foliage, protected by moats, inhabited by an emperor who is never seen, which is to say, literally, by no one knows who. Daily, in their rapid, energetic, bullet-like trajectories, the taxis avoid this circle, whose low crest, the visible form of invisibility, hides the sacred "nothing". One of the most powerful cities of modernity is thereby built around an opaque ring of walls, streams, roofs, and trees whose own center is no more than an evaporated notion, subsisting here, not in order to irradiate power, but to give to the entire urban movement the support of its central emptiness, forcing the traffic to make a perpetual detour. In this manner, we are told, the system of the imaginary is spread circularly, by detours and returns the length of an empty subject.
>
> (Barthes, 1982, p. 30)

Tokyo's fragmented identity was investigated by the architect Yoshinobu Ashihara[4] (1918–2003), who argued that the city's chaotic exterior conceals a hidden order, an "aesthetics of chaos": "Its chaotic layout, fascinating if somehow inscrutable, is the result of an orientation to space quite different from that which governs the architecture, townscapes, and city planning of the West" (Ashihara, 1989, p. 13). Tokyo, according to Ashihara, is a city that is perennially in flow, an "organic amoeba city": "The huge megalopolis of Tokyo and, similarly, other large Japanese cities illustrate this best: their endless sprawl encroaches on the surrounding countryside like a silkworm munching at a mulberry leaf. As a whole, the shape of these cities is extremely unstable and undefined, for their fringes are surrounded by an intermediary, ambiguous zone that is in constant flux" (ibid., p. 56).

The *flâneur*

Approaching the city as that place in which "even if you lose your way, you cannot go wrong" (Abe, 2001) brings us face to face with the contemporary metropolis, which allows of infinite movement and yet can never be fully circumscribed. The city of today has become a literary place of wandering about the streets, viewed as "the dwelling place of the collective" (Benjamin, 1999, p. 423), while the labyrinth is used to represent getting lost in the urban context: "The labyrinth is the right path for him who always arrives early at his destination. For the *flâneur*, this destination is the marketplace" (ibid., p. 338). Thus, the figure of the *flâneur*, produced when cities were buoyed up by the confidence of the modern era, is one who wanders about the metropolis, refusing to engage with its rigidly planned layout. And the powerful metaphor of the labyrinth recurs again and again in artistic and literary explorations of the city. Both the city of the twentieth-century *flâneur*, and the global city of today clearly converge upon it.

Paris "around 1900": "the realization of that ancient human dream, the labyrinth. It is this reality to which the *flâneur*, without knowing it, devotes himself" (ibid., pp. 430–431). The *flâneur* dawdles randomly among the crowds: "The crowd is his domain, as the air is that of the bird and the water that of the fish. His passion and his profession is 'to wed the crowd'. For the perfect *flâneur*, for the passionate observer, it is an immense pleasure to choose his home in number, change, motion, in the fleeting and the infinite" (Baudelaire, 2008, p. 9). For Benjamin, the "masses in Baudelaire" "stretch before the *flâneur* as a veil: they are the newest drug for the solitary. Second, they efface all traces of the individual: they are the newest asylum for the reprobate and the prescript. Finally, within the labyrinth of the city, the masses are the newest and most impenetrable labyrinth" (Benjamin, 1999, p. 446). The *flâneur* reads urban space as a text and loses himself within it:

"Not to find one's way around a city does not mean much", wrote Benjamin in 1932:

> But to lose one's way in a city, as one loses one's way in a forest, requires some schooling. Street names must speak to the urban wanderer like the snapping of dry twigs, and little streets in the heart of the city must reflect the times of day, for him, as clearly as a mountain valley. This art I acquired rather late in life; it fulfilled a dream, of which the first traces were labyrinths on the blotting papers in my school notebooks.
>
> (Benjamin, 2006, p. 352)

The aimless wandering of the *flâneur*[5] – a figure that has returned to the fore in what Nuvolati refers to as our "late-modern society" (Nuvolati, 2006,

p. 7) – implies rejection of a model of city that has become indifferent to the concerns of "ordinary citizens, those who do not have the power to get their hands on their city and change the face of the environment in which they live" (La Cecla, 2000, p. 38). A situation that is typical of the world city of today, a city of security cameras and shopping malls, and public spaces to be defended from "improper use" by tramps, immigrants, children, and animals. La Cecla goes on to comment that:

> It was against this terrible dominion over place that city police regulations and urban planning first came into being. The city had to be cleared of its local inhabitants and made permeable and controllable from the outside, in its entirety. A "clean sweep" was to be made, by demolishing, evicting, and penetrating courts and courtyards [...] The city must cease to be a set of forests in which strangers lose their way.
>
> (ibid., p. 33)

In the late nineteenth century, the meandering of the *flâneur* also expressed criticism of Second Empire society and the notion of modernity: Benjamin criticized the ongoing transformation of the city according to rational and functionalist criteria, seeking out "pockets of resistance" among the traditional marketplaces and districts. The street can become an "interior": "For if *flânerie* can transform Paris into one great interior, a house whose rooms are the *quartiers*, no less clearly demarcated by thresholds than are real rooms, then on the other hand the city can appear to someone walking through it to be without thresholds" (Benjamin, 1999, p. 422).

Wandering, from the Dadaists to the Situationists

In the 1920s, the *flâneur*'s art of wandering mutated into "excursions" led by the Dadaist movement to: "chosen places, particularly those that have no reason to exist", as stated in a pamphlet distributed by Dadaists on the streets of Paris in 1921 (Dada à Saint-Jullien-le-Pauvre, 14 April 1921). The idea was to wander about the city, exploring its "trivial places" in search of mental associations. In Careri's view: "Dada elevated the tradition of *flânerie* to an aesthetic operation" (Careri, 2006, p. 47). Next came the "deambulations" of the Surrealists, conceived as a means of exploring "the boundaries between waking life and the dream life" (Breton, 1973, pp. 53–54). The experience of wandering about city streets was also described by Louis Aragon, in *Paris Peasant*: "How oddly this light suffuses the covered arcades which abound in Paris in the vicinity of the main boulevards and which are rather disturbingly named *passages*, as though no one had the right to linger for more than an instant in those sunless corridors" (Aragon, 1994, p. 28).

In the 1950s, the Surrealist approach was taken up again by proponents of Situationism: Guy Debord revisited the practice of urban wandering, labelling it "psychogeographical drifting" (in French "dérive") to distinguish it from the experience of making a journey or going on a walk. In the words of Careri: "Urban drifting consists of intentionally losing oneself among the folds of a city" (Careri, 2001, p. 32). This practice was proposed as a method of reading the city according to the principles of "psycho-geography", a discipline that acknowledges "the psychic effects of the urban context on individuals and their behaviour" (Constant in Careri, 2001, p. 27). The practice of drifting also led to the production of a cartographic apparatus that was presented at the "First Exhibition of Psychogeography" in 1957, in an attempt to deconstruct traditional cartography: psychogeographical maps (*Guide psychogéographique de Paris; The Naked City. Illustration de l'hypothèse des plaques tournantes en psychogéographie*) are maps that invite the user to get lost, using cut-out sections of the traditional map of Paris to represent the city as a fragmented and discontinuous space. A system of arrows connects and separates the different segments, each of which represents a homogeneous area with its own *unity of atmosphere*, as established via psychogeographical observation: thus, the "naked city" is revealed, that which Debord and Asger Jorn set out on their driftings to find, a city that is imbued with relations of attraction, repulsion, or exclusion among the different areas. These relations were based on different memories of events such as war and colonialism, and were expressed through texts and postage stamps amongst other channels. The Situationists[6] observed the city from the outside, living out the sensory, ludic, and affective experiences it offered, and ultimately developing a radical critique of contemporary society: "This society which eliminates geographical distance reproduces distance internally as spectacular separation" (Debord, 1967, p. 167).

The city on which all these observations were based was Paris, *la ville qui remue* or "the city that never stops moving", the city *par excellence* in terms of architecture and culture. A city that was under threat from new urban planning projects wiping out its small streets, les Halles and its *Passages* (arcades). Debord argued that:

> The concern to have open spaces allowing for the rapid circulation of troops and the use of artillery against insurrections was at the origin of the urban renewal plan adopted by the Second Empire. But from any standpoint other than that of police control, Haussmann's Paris is a city built by an idiot, full of sound and fury, signifying nothing.
>
> (Debord, 1955a, p. 13)

The drift was therefore in the direction of a resistance strategy conducted among what was left of the old quarters and their inhabitants "whereby losing oneself was a practice of restoring value to the undecidable and radically anarchic aspect of spatial experience" (La Cecla, 2000, p. 132). Debord went to the lengths of listing the best places and worst places to visit from this point of view:

> Urbanism. In Paris, we currently recommend spending time in: "Continent" Contrescarpe; Chinatown; the Jewish Quarter; Butte-aux-Cailles (the labyrinth); Aubervilliers (by night); the squares of the 7th Arrondissement; the Institut Médico-Légal; Rue Dauphine (Nesles); Buttes-Chaumont (play); Quartier Merri; Park Monceau; Île Louis (the island); Pigalle; les Halles (Rue Denis, Rue du Jour); Quartier de l'Europe (memory); Rue Sauvage. We recommend avoiding at all costs: the 6th and 15th Arrondissements; the large boulevards; the Luxembourg Garden; the Champs-Élysées; Place Blanche; Montmartre; the École Militaire; Place de la République, Place de l'Étoile and Place de l'Opéra; the entire 16th Arrondissement.
>
> (Debord, 1955b, p. 1)

Producing psychogeographical maps with the power "to turn the whole of life into an exciting game" requires following a logic of insubordination. In elaborating on this point Debord related that: "A friend recently told me that he had just wandered through the Harz region of Germany while blindly following the directions of a map of London. This sort of game is obviously only a feeble beginning in comparison to the complete creation of architecture and urbanism that will someday be within the power of everyone" (Debord, 1955a, p. 5).

Debord's critique of the "society of the spectacle" anticipated contemporary reflection on urban disorientation in cities that are becoming increasingly vast and frenetic: "The self-destruction of the urban environment is already well underway. The explosion of cities into the countryside, covering it with what Mumford calls 'a formless mass of thinly-spread semi-urban tissue', is directly governed by the imperatives of consumption" (Debord, 1994, p. 174). Debord condemned the dictatorship of the automobile, which had been the "pilot" consumer good, the dominance of motorways which he attributed with disrupting the local territory, and finally shopping centres, those true "temples of frenetic consumption". He viewed the "society of the spectacle" as the culprit for the deteriorating quality of life in cities: "But the technical organization of consumption is only the most visible aspect of the general process of decomposition that has brought the city to the point of consuming itself" (ibid.). However, the Situationists did not limit themselves to formulating critiques, but also explored and designed new and alternative spaces. In 1956, Constant Nieuwenhuys visited a gypsy camp in Alba on the banks of the Italian river Tanaro, an experience that brought him into contact with nomadic culture and new forms of dwelling. This ultimately led him to develop the concept of a New Babylon "a permanent encampment. [...] a New Babylon where, under one roof, with the aid of moveable elements, a shared residence is built; a temporary constantly remodeled living area; a camp for nomads on a planetary scale" (Constant in Careri, 2001, p. 26).

New Babylon was conceived by Constant as a utopia put into practice, a revolutionary project that radically critiqued bourgeois-utilitarian society, proposing in its place a playful society in which aggression would yield to creativity:[7] "New Babylon is a ludic city, a collective work that is built thanks to the architectural creativity of a new wandering society, a population that infinitely constructs and reconstructs its own labyrinth within a new and artificial landscape" (Careri, 2006, p. 78). New Babylon would be a fluctuating city, thus facilitating chance contact and encounters: "New Babylon ends nowhere (since the earth is round); it knows no frontiers (since there are no more national economies) or collectivities (since humanity is fluctuating). Every place is accessible to one and all. The whole earth becomes home to its owners" (Constant, in Careri, 2006, p. 84). In the "dérive" of the post-war period, the city was imagined as a nomadic, mobile, self-organizing space, and this "New Babylon" anticipated the web of the global era: a contemporary practice of drifting, we might say of cyber-drifting, is wandering about the satellite network, and losing oneself among the streets of Google Maps. Thus, getting lost in the city-labyrinth takes on a positive value: in the words of Careri "Going astray ['*errare*'] may be viewed as a *value* rather than as a mistake [*errore*]" (Careri, 2006, p. 9).

A world of citizens

Pantopolis, world city, ecumenopolis, cosmopolis, global city: this rich lexicon reflects the interest in the theme of urban transformation in the globalization era, as well as the questions and concerns – both quantitative and qualitative in nature – that are currently being prompted by the growth of the urban phenomenon. In Oswald Spengler's view:

> World history is city-history. […] Finally, there arises the monstrous symbol and vessel of the completely emancipated intellect, the world-city, the centre in which the course of a world-history ends by winding itself up. […] The stone Colossus, "Cosmopolis", stands at the end of the life-course of every great Culture. […] Looking down from one of the old towers upon the sea of houses, we perceive in this petrification of a historic being the exact epoch that marks the end of organic growth and the beginning of an inorganic unrestrained process of agglomeration.
>
> (Spengler, 1991, pp. 247–48)

But how may we define the city of today? In 2008, for the first time in history, the world's urban population overtook the number of rural residents, as reported by the UN in the *State of World Population 2007: Unleashing the Potential of Urban Growth*: "an invisible but momentous milestone" for humanity at the dawn of the third millennium. Today, this population comprises some 3.3 billion people, a figure that is predicted to rise to around 5 billion by 2030 (UNFPA, 2007, p. 1). The city is expanding, with the megalopolis dominating the global scene. Nonetheless, urban growth – in both spatial and temporal terms – has failed to meet the millenarian expectations of the last century:

today, megacities are home to only 4 per cent of the world's overall population and 9 per cent of its city-dwellers, and this pattern is not predicted to change in the near future (UNFPA, 2007, p. 9). The growth in land area occupied by cities has also been relatively modest. In this regard, the UNFPA report states that: "The space taken up by urban localities is increasing faster than the urban population itself. Between 2000 and 2030, the world's urban population is expected to increase by 72 per cent, while the built-up areas of cities of 100,000 people or more could increase by 175 per cent" (UNFPA, 2007, p. 45). The expansion of the city occurs in waves, following the rhythm of the tide: "Rhythms imply repetitions and can be defined as movements and differences in repetition" (Lefebvre, 2004, p. 231).[8] Based on developments to date, how might we envisage the city of the future? Can we trust projections of an urban agglomeration in continuous expansion, suggesting that we will end up with a pantopolis sprawled across the world's continents? Currently, the area taken up by cities is not as vast as we might imagine, considering that over half the world's population reside in it. Again, UNFPA data based on satellite observations shows that urban settlements, including both built-up areas and green spaces, take up no more than 2.8 per cent of the Earth's land surface: this implies that 3.3 billion people occupy an area that is less than half the size of Australia (UNFPA, 2007).

Cities are growing in number and size, but the urban phenomenon remains a network of economic, political, and human relations with increasingly densely woven connections and fewer spaces. Bauman has argued that:

> Contemporary cities are the battlegrounds on which global powers and stubbornly local meanings and identities meet, clash, struggle and seek a satisfactory, or just bearable, settlement – a mode of cohabitation that is hoped to be a lasting peace but as a rule proves to be but an armistice, an interval to repair broken defences and redeploy fighting units. It is that confrontation, and not any single factor, that sets in motion and guides the dynamics of the "liquid modern" city.
>
> (Bauman, 2003, p. 21)

The city is therefore not only a physical object, but rather a worldview that is currently shared by the majority of the world's inhabitants; it is a cultural product that ubiquitously spreads its own distinctive lifestyle and mode of internal and external relations. Thus, Capel has argued that: "World city means, in some sense, a world system of cities, a development that has been examined, especially since historians and economists first clearly invoked a world socioeconomic system, from Marxist perspectives based on theories of social change or other perspectives emphasizing the 'globalization' of the economy" (Capel, 2003, p. 46).

The city is the key place for markets, information flows, fashion and cultural trends; at the same time, it is the place of overcrowding, detachment from nature, traffic, and pollution. Above all it is the place of the crowd, of the growing number of people caught up in an urban lifestyle. In the city of the future, the majority of citizens will be poor; according to the UNFPA 2007 document: "One of the Report's key observations is that poor people will make up a large part of future urban growth. This simple fact has generally been overlooked, at great cost. Most urban growth now stems from natural increase (more births than deaths) rather than migration. But wherever it comes from, the growth of urban areas includes huge numbers of poor people" (UNFPA, 2007, p. 1). In the rapidly growing global urban population, the majority of citizens will be from the countries of the Global South: the urban populations of Africa and Asia are forecast to double between 2000 and 2030. UNFPA predicts that in 2030, the cities of the Global South will be home to 80 per cent of the world's urban population (UNFPA, 2007).

Urban space is growing, spreading itself out over interminable shantytowns, green spaces, megalopolitan areas, networks of cities, and suburbs, to form a tissue that at times is tightly knit and at times fragmented. Indeed, in relation to the work of Lefebvre, Mike Crang has argued that "instead of being a solid thing, the city is a becoming, through circulation, combination and recombination of people and things. This is a seductive vision where the urban field becomes an object in motion, or rather an object with time" (Crang, 2001, p. 190). Thus, space, as the concept underpinning disorientation, takes on a temporal dimension. The city, as described by Henri Lefebvre, lives its days and nights according to its own particular rhythms: "The *multiple* rhythms of modernity: the various speeds of circulation; the different spacings of movement; and the varied directions of flows" (Highmore, 2005, p. 11). For Lefebvre, the rhythm of the urban centre is consequent on the body and its ability to move: "Everywhere where there is interaction between a place, a time, and an expenditure of energy, there is rhythm" (Lefebvre, 2004, p. 15). In analyzing this dimension of city life, he distinguishes between the rhythms of Mediterranean cities – Barcelona, Beirut, Naples, Marseilles, or Tunis – and those of towns on the Atlantic coast: "The latter are governed by the cosmic rhythms of the tides – lunar rhythms! As for Mediterranean cities, they skirt a sea with almost no tide and therefore the cyclical time of the sun takes on a predominant importance. Lunar cities of the ocean? Solar cities of the Mediterranean? Why not?" (Lefebvre, 1995, pp. 231–232). From the window of his Parisian apartment in Rue Rambuteau, opposite the Beaubourg quarter, Lefebvre listened to the rhythm of his neighbourhood, and invited his readers to attend to the pauses and movements of people and vehicles, and the different tonalities of urban rhythm, as he heard them from his vantage point at the window: "One does not chatter while crossing a dangerous junction under the threat of wild cats and elephants ready to charge forward, taxis, buses, lorries, various cars. Hence the relative silence in this crowd. A kind of soft murmuring, sometimes a cry, a call" (ibid., p. 220).

The contemporary city[9] becomes a polyrhythmic ensemble (Virilio, 1997) for those who live, work, and circulate in it and walk its streets, a diffused city. Capel: "At the same time, the constructed city (urbs) diffuses and the diffused or dispersed city reflects this state of affairs. Alongside the compact city, today increasingly valued, there is the dispersed and diffused city, with instant connections via telephone networks and mobile phone access" (Capel, 2003, p. 48). How does one get lost in the "diffused city"? How may its full parts, the solids, and its empty parts, the voids, be recognized and experienced? In the literary imagination, just as one can get lost in Kobo Abe's Tokyo and Paul Auster's New York, so disorientation features strongly in Italo Calvino's *Invisible Cities.* Among the many cities visited by Marco Polo, according to Calvino, we can identify the city of Trude, which looks identical to any other, with the same kind of houses and streets, giving first-time visitors the feeling that they have already been to this city, already arrived at this same airport, already stayed at this same hotel. Indeed, why travel to Trude in the first place, given that its sameness makes one want to leave again? One can immediately board another airplane to get away: "But you will arrive at another Trude, absolutely the same, [...] The world is covered by a sole Trude which does not begin and does not end" (Calvino, 1974, p. 128). The city of *urban sprawl* and "informal neighbourhoods" no longer corresponds to an organic entity or project, but is essentially a network of virtual relationships that is fluid and changeable, and lacks a centre, map, or coordinates:

> These cities are configured in a topology where all places are equivalent to each other, where every citizen lives their own city, which might have nothing to do with their fellow citizen's experience. The new centrality, if there is such a thing, is given by the shopping malls, the multi-purpose spaces where businesses have brought back together what the city broke up.
>
> (Rogel Salazar, 2003, p. 13)

The urban rhizome

The dominant metaphor of today's city is therefore once again the labyrinth, albeit a labyrinth that takes on new spatio-temporal forms. A phenomenon early detected by Debord: "In architecture itself, a taste for the dérive tends to sanction all sorts of new forms of labyrinth, which are facilitated by modern architectural possibilities" (Debord, 1956, p. 23). The labyrinth takes the form of a rhizome (see § 7.1. and §10.3), of a new discovery, or unforeseen encounter: "The rhizomatic approach allows us to understand the logics of different contemporary cities, the ones with dimensions never thought of, that are now broken in several 'senses'. They are cities without unity, since they contain all sorts of spaces from different times and models" (Rogel Salazar, 2003, p. 13). The rhizomatic labyrinth helps us to engage with the complexity of contemporary urban space, and with the new logics and problems of today's city: a city of unprecedented dimensions, with multiple centres, whose citizens may live as total strangers alongside other citizens.

Rogel Salazar elaborates on the theme: "The labyrinth is an icon of perplexity, where it is possible to feel the anguish and vertigo symbolised by the transgression of the space-time. And even if the labyrinth creates disorientation and makes references to infinity, its design allows us to see certain partial and occasional order immersed in an apparent nonsense" (ibid., p. 16). The metaphor of the labyrinth thus becomes a cartographic strategy for finding our way through the complex present-day city, exploring it, getting to know its spaces, overcoming our fears of it, and developing ways of coping with our sense of spatial disorientation:

> The new spatial logic, characteristic of the Informational City, is determined by the pre-eminence of the space of flows over the space of places. [...] Urban life muddles through the pace of history. When this pace accelerates, cities – and their people – become confused, spaces turn threatening, and meaning escapes from experience. In such disconcerting yet magnificent times, knowledge becomes the only source to restore meaning, and thus meaningful action.
>
> (Castells, 2006, p. 136)

In the rhizomatic labyrinth, the city is not only at the heart of the economy, politics, and the culture of modernity: we also find the city produced by the digital revolution, the city of networks, information highways, shopping malls, which have become – like the Parisian arcades for the flâneur – the refuge of the cyberflâneur. As observed by Castells: "The information age is ushering in a new urban form, the informational city [...] because of the nature of the new society, organized around networks, and partly made up of flows, the informational city is not a form but a process, a process characterized by the structural domination of the space of flows" (Castells, 1996, p. 398). Cyberflâneurs do not only wander about large shopping centres and department stores; they enjoy unlimited global mobility which cancels out physical space and allows them to move about the network of the virtual cosmospolis, with its potential for interaction and its capacity to redefine the systems with which it enters into contact: we find ourselves in the hyperlabyrinth of the modern city in which technology is inevitably called in to restore equilibrium when under threat from any kind of crisis (Raffestin, undated). However, the hypertechnological city of the information era does not erase or exclude a geographical sense of place. The labyrinth that is the World Wide Web does not only offer passive images, like at the cinema or on television, but also the opportunity to interact with images and their producers; on computers, mobile phones, and digital cameras, the virtual and real worlds coexist and mutually influence one another: an example is GPS navigation devices in cars with their extraordinary power to orient or disorient (if not regularly updated) our movements through the urban labyrinth. In cyberspace (see Chapter 10): "instantaneous connections are possible which render physical spatial differences irrelevant" (Featherstone 1998, p. 921).

The mongrel city

The metaphor of the labyrinth represents diversity, the cultural disorientation of the newly arrived, having to adapt to new ways of life, new forms of communication, and new survival strategies. On this theme, Saskia Sassen observed that:

> The large Western city of today concentrates diversity. Its spaces are inscribed with the dominant corporate culture but also with a multiplicity of other cultures and identities. The slippage is evident: the dominant culture can encompass only part of the city. And while corporate power inscribes these cultures and identifies them with "otherness" thereby devaluing them, they are present everywhere.
>
> (Sassen, 1998, p. xxxi)

Urban space is imbued with diversity, which may be expressed in both formal and informal architectural forms. Global lifestyles are forged with residues of local knowledge, producing new ways of being and of attributing meaning to the surrounding environment, but also new sources of anxiety and aggression. We are confronted with the political paradox described by Bauman: "*Cities have become dumping grounds for globally begotten problems.* The residents of cities and their elected representatives tend to be confronted with a task they can by no stretch of imagination fulfil: the task of finding local solutions to global contradictions" (Bauman, 2013, p. 101, italics in the original). City-dwellers are switching to a "glocal" rhythm (Swyngedouw, 1997; Bauman, 1998).

The world city is not only a functional machine that generates and manages capital, information, politics, and people in a globally connected market, it is also a space in which different types of knowledge and different cultures meet, clash, and merge. The vertiginous growth of cities is underpinned by the arrival of multiple generations of migrants, from a variety of social and cultural backgrounds (see § 3.8–3.9). Thus, the cosmopolis is a multiform universe that moves to the rhythms of different cultural actors and multiple ethnic identities. It is, in the eyes of Leonie Sandercock: "The city of memory, of desire, of spirit: the unruly city as opposed to the planners' dream of the rational city – a dream that came out of the social sciences with its social and spatial technologies of control" (Sandercock, 1998, p. 207). The inhabitants of the new cities are bombarded with new stimuli, new opportunities, and new difficulties across the professional, familial, and recreational areas of their lives. The city is the place of cultural enrichment, exchange, and encounter, but also the place where one can lose one's community ties, traditional values, sense of territorial belonging, and contact with nature: the new urban dwellers are often disadvantaged, suffering from marginalization, exclusion, and displacement,[10] which in turn can lead them to engage in violent and aggressive behaviours. Furthermore, the way of life in global cities is often associated with the imposition of Western values and lifestyles, and this can be a source of resentment (UNFPA, 2007).

The different rhythms of the city develop out of the movements and encounters of a heterogeneous population that continuously produces new cross-bred identities and new forms of expression. This gives rise to "the mongrel cities of the twenty-first century" as they have been described by Sandercock:

> I have appropriated "mongrel" from Salman Rushdie (1991) and use it in the same spirit as he does: as both provocation and term of approbation for the kinds of changes that are happening in cities the world over, as they become more multi-ethnic and multi-cultural. This book is a manifesto of sorts: a radical manifesto for twenty-first century cities.
>
> (Sandercock, 2003, p. XIII)

In the "mongrel" scenario described by Sandercock, the planetary citizen is required to cope with new forms of disorientation that span all life spheres, in a melting pot of stimuli and diversity. In reading the city through the eyes of its new citizens, Sandercock collected and analyzed the stories of migrants from different cultural backgrounds (see Chapter 5). The world city confronts us with new identities and new forms of citizenship that call for: "the imagination and knowledge required to leave the stagnant waters of the monoculture and acquire a deep appreciation of the advantages of mixing, cross-breeding and fusing diversity" (Agostinis, 1996, p. 202). The inhabitants of a given city may not share the same origins, cultural and political traditions, and background experiences: "Yet", as Featherstone has argued, "in contrast to the polis, this cosmopolis possesses a tolerance of diversity" (Featherstone, 1998, p. 911), enabling new "cosmopolitan identifications".

However, this "mongrel city" of chaos and diversity is not only the legacy of "underdevelopment", to be tidied up and reorganized to endow it with the rational order lacking in Barthes' Tokyo. The city's key role as a financial, political, economic, and tourism hub is also invoked today by those seeking to impose a single, organized, and planned model on the pantopolis; Caterina Resta has speculated as to the possible endpoint of such efforts:

> The cosmopolis therefore becomes the World State[11] of which all are citizens but at the cost of completely cancelling out all singular idioms, languages, communities, and memories. It is no coincidence that the modern nomad "communicates" via the only universal language, that of technology, which on its part breaks up the ancient stones by subjecting the earth and humankind to incessant labour.
>
> (Resta, 1999, p.82)

However, despite the technological advances achieved to date, today's city has not yet developed the ability to communicate through a single language nor to forgo its diversity. According to Eichberg: "The movement culture of the city appears as a multiplicity of highly different movement cultural 'scenes' [...]. Indeed, the metropolis reveals itself as an archipelago of networks, places and milieus, some interacting with each other, others non-connected – a world of islands and channels" (Eichberg, 1998, p. 277). Contemporary cities are lived, travelled, loved, hated, and constructed thanks to the hands, gazes, bodies, and ideas of some 3.5 billion people: clearly therefore, the city cannot but be the place of heterogeneity.

Suburbia (the city below)

The city expands both horizontally and vertically. The city of the skyscrapers merges with the many levels of the underground city. We thus enter the "suburbia", in which the city's inhabitants live and function. In 1968, Robert Smithson, an artist and exponent of the Land art movement, had this to say:

> Suburbia encompasses the large cities and dislocates the "country". Suburbia literally means a "city below"; it is a circular gulf between city and country – a place where buildings seem to sink away from one's vision – buildings fall back into sprawling babels or limbos. Every site glides away toward absence. An immense negative entity of formlessness displaces the centre which is the city and swamps the country.
>
> (Smithson, 1979, p. 74)

The subterranean city is metaphoric and real at the same time. Underground canals, sewage systems, and abandoned military bases can provide shelter to marginalized populations,[12] while in the urban planning of many contemporary cities, underground passageways can become the site of sophisticated shopping malls and leisure centres, as in Montreal's Indoor City. The city's subterranean labyrinth includes its underground transport network, a place of avant-garde technology and, at the same time, a potentially disorienting space (see § 2.2). The metropolitan subway can therefore become a symbolic site, in which modernity and tradition, centre and periphery, orientation and disorientation are in contrast with one another.[13] Rachid Boudjedra, in his novel *Topographie idéale pour une aggression caractérisée*, narrates the odyssey of an Algerian immigrant, who is newly arrived in Paris and unable to speak or write French, and his experience of disorientation in the underground. The book, which is not divided into chapters but by metro lines, traces the migrant's wanderings through the subterranean labyrinth from when he leaves the Gare d'Austerlitz, with a heavy suitcase in his left hand and, in his right, a piece of paper with his cousin's address on it (written out for him by a little girl in his home village). Lost among symbols, colours, and signs that are impossible for him to decipher, the protagonist rambles on through the meanders of the underground network, unable to identify meaningful landmarks, and disoriented both spatially – as he tries to find his way among metro lines represented in different colours and in an unintelligible alphabet – and mentally – due to an overload of information, incomprehensible messages, and alien characters. His confusion is such that he ends up: "not knowing the north from the south [...] the east from the west" (Boudjedra, 1975, p. 19).

The nameless migrant feels under attack from the lights and the immodest and incomprehensible advertising, the escalators, and the crowd: "this aerial topography disturbed him, [...] these doublings of spaces [...] appearing from anywhere perpendicularly, in parallel, vertically, horizontally, [...] deeply traumatizing him [...] like a mouse caught in a maze" (ibid., p. 115). After an entire day of underground wanderings, "always continuing on his hallucinatory marathon, turning around, bumping into people" (ibid. p. 188), he finally manages to reach his destination on the outskirts of Clichy, only to be brutally murdered by a gang of neo-Nazi skinheads as he surfaces from the underground station.

The novel is based on a real-life murder, one of a series of racist attacks perpetrated in France in the years 1971–1975. According to Rosello: "*Topographie* is a performative textual account of the migrant's constant disorientation" (Rosello, 2016, p. 3). However, I do not agree with Rosello that Boudjedra's style of writing – long sentences, lack of punctuation, endless repetition – makes "This novel [...] almost unreadable". Or that "this is the kind of book that you would rather, at some point, abandon" (ibid.). On the contrary, in my opinion it is a profoundly moving story, that right to the end successfully transmits to the reader, the disoriented protagonist's emotions, contradictions, and perceptions of an increasingly hostile space.

Seeing and making the city

As we wander through the rhizomatic labyrinth of the city, we run up against the double level of meaning that is inherent in the city, as pointed up by Michel de Certeau in *The Practice of Everyday Life*: the city of "seeing" and the city of "doing". In the "panorama-city", urban space may be viewed from above, transposed onto a map, planned, and organized. From this vantage point, de Certeau proposed: "Escaping the imaginary totalizations produced by the eye, the everyday has a certain strangeness that does not surface, or whose surface is only its upper limit, outlining itself against the visible" (de Certeau, 1984, p. 93). However, on descending to street level, one becomes a walker or pedestrian: "A *migrational*, or metaphorical, city thus slips into the clear text of the planned and readable city" (ibid., italics in the original). Walkers are blind in that they cannot view the city as a unified whole, but experience the everyday through their spatial and social relations. In de Certeau's own words:

> The ordinary practitioners of the city live "down below", below the thresholds at which visibility begins. They walk – an elementary form of

> this experience of the city; they are walkers, *Wandersmänner*, whose bodies follow the thicks and thins of an urban "text" they write without being able to read. These practitioners make use of spaces that cannot be seen; their knowledge of them is as blind as that of lovers in each other's arms. [...] It is as though the practices organizing a bustling city were characterized by their blindness.
>
> (ibid.)

The *walker*'s city is the liveliest and most creative part of urban space: it tells the stories of millions of individuals who dwell in this space, and their movements along different itineraries; it is rich with open spaces, dark corners, new ways of being and of relating to the outside and the inside, experiences of getting lost and of finding oneself again: "A mutilation that is widely practiced in a capacity that is still alive and possible, and of which we still find evidence all round us. [...] the capacity for dwelling is so closely bound up with a given culture that the 'local mind' is able to organize and express itself even in the most miserable urban grid or the most squalid slum. And indeed the barriadas, favelas, and bidonvilles, the urban ghettos of the marginalized are more alive than the 'smart' areas of the city" (La Cecla, 2000, p. 5). The underground city, viewed from above, may seem chaotic and aggressive, but for those who live there it is home, a familiar, accessible, and recognizable place: "The urban maze is a picture of the stranger's movement in an estranged world. It is a picture of alienation in the urban labyrinth as seen from outside. For the inhabitants of the city, the urban space offers a labyrinthine pattern of feeling at home. [...]. Seen from inside, the urban labyrinth is a picture of identity" (Eichberg, 1998, p. 78).

In the "shadow cities" or the "peripheries of the soul" (Agostinis, 1996), a billion squatters devote themselves to producing the city: often multicultural and foreign citizens, they are both actors in and producers of the space around them: "The many displaced, deterritorialized and transient populations that constitute today's ethnoscapes are engaged in the construction of locality, as a structure of feeling, often in the face of the erosion, dispersal, and implosion of neighbourhoods as coherent social formations" (Appadurai, 1996, p. 199). The appropriation of space in the so-called *terrains vagues*, the no man's land on the abandoned outskirts of cities, produces new forms of dwelling and even a new sense of place: "It is the very nature of lost places and places of losing oneself that makes marginalized areas and wastelands, the spaces par excellence in which the subjects of the local mind can be fully active" (La Cecla, 2000, p. 138). The inhabitant of *slums* often build the city themselves, a city that has its own rules, power relations, and market dynamics. In Neuwirth's view: "The miracle is that the world's squatters value civil society and want to find a way of working within the system. They are law-abiding outlaws, patriotic criminals" (Neuwirth, 2005, p. 310).

Practices of resistance

Our flâneurs, now become cyberflâneurs, wander about the pantopolis. They have not adjusted to the frenetic pace of the new global metropolis, on the contrary, it often makes them suffer a sense of displacement and loneliness:

> Hence the similarities often proposed in the sociological and non-sociological literature, between the flâneur and those who are marginalized within society: the homeless person who wanders about the railway stations, the retired person who is always chasing memories or obsessively monitoring the progress of any roadworks being carried out in the area, the immigrant who has moved to a strange city in search of a new identity, the shanty-dweller who spends his or her day walking about the city centre, the poor student whiling away time in places of leisure.
>
> (Nuvolati, 2006, p. 33)

The construction of urban space by popular cultures is in conflict with the dominant ideology of the "new barbarians who invaded the city with a view to turning it into a Global Duckburg that resides in detached family homes and extends its habitat along the real motorways and the virtual highways of Internet" (Careri, 2006, p. 130). The new meanings and values produced are often in contrast with the official ones laid down by the hegemonic sectors of urban society. Just as the flâneur of the last century opposed the modern city, the virtual flâneur practices a strategy of resistance to the tastes, values, and ideologies of the pantopolis. The practice of "transurbance" is currently implemented by the Roman laboratory, Stalker (Careri, 2006) which, as its name suggests, is literally composed of "individuals who walk slowly and stealthily" (Nuvolati, 2006, p. 71). Stalker organized their first walk through "alternative areas" in 1995 and since then have continued to organize nomadic exploration experiences and walking as an aesthetic practice, both in Italy and abroad.[14]

The renewed appeal of the labyrinth and urban nomadism is confirmed by the sudden and unforeseen appearance of undecipherable writing on city walls; from political messages rejecting society and its institutions; to the artistic language of suburban graffiti and murals, and the affirmation of the work of graffiti writers on the streets: "The global city is a strategic place for actors without power, since it enables them to state their presence" (Sassen, 2001, p. 66). This city conceals a complex and dynamic scene, a mosaic of diverse and fragmented spaces in which it is still possible to get lost among thousands of possible paths:

> The new labyrinthism is confirmed by the new scripture on the urban walls, by the language of graffiti. [...] Unreadable names appear on the concrete. By winding lines on the walls, the graffiti painter does not formulate any appeal to action. Rather, the graffiti marks a claim of identity: "This is me!" – "I was here!" – "This is my place!" This is a new

> variation of the quest of identity in the labyrinth, the question – "where and who am I?"
>
> (Eichberg, 1996, p. 15)

In the chapter "Heretical Geography 1: The Crucial 'Where' of Graffiti"[15] (see § 10.7), Tim Cresswell examined the political significance of early graffiti in New York, from the 1970s onwards, and the process of its "strange transformation into art" (Cresswell, 1996, p. 35) from its previous status of "dirt, garbage, pollution, obscenity" (ibid. p. 39). For Cresswell, graffiti represents a political and moral challenge, with anarchist overtones, to the city's established order:

> Graffiti is continually portrayed [...] as the chaotic untamed voice of the irrational. As such it is both resisted and condemned. Graffiti is rebellious, irrational, dirty, and irreducibly "other". In these senses it is connected to the third world and to immigrants – themselves described as rebellious and irrational. [...] The graffiti artist (like the rebellious third world) is the insane spoiler who resists reason and introduces chaos.
>
> (ibid., p. 45)

The presence of graffiti on city walls and the political battle to remove it, also call into question the concepts of sense of belonging, the "appropriate" use of places and the "dominant dichotomy" between public and private space: "It interrupts the familiar boundaries of the public and the private by declaring the public private and the private public. Graffiti appears on the streets, the facades, the exteriors, and the interiors that construct and articulate the meanings of the city. To the graffiti writer everywhere is free space" (ibid., p. 47). The debate about graffiti and wall art, their value, and artistic, educational, and decorative functions is ongoing in many cultural contexts, producing different interpretations and meanings and a continuous stream of new graphic and expressive styles. There is a great difference for example in the language and interpretation of wall paintings in South African cities during and after apartheid, seen as fulfilling a social and educational function, but also as a form of condemnation, racial empowerment, and memory of oppression (Antheaume and Deliry-Antheaume, 1999).

The presence of the cyberflâneur is thus borne out by attempts to attribute new meanings to the city's more soulless places and to reclaim urban spaces, contravening their conventional and sanctioned uses. Examples include city parks, gardens, and squares occupied by immigrant communities or groups of adolescents; the filing through streets of rallies, protests, and parades; the places of rap, breakdancing, alternative music, and rave parties. Invoking the principle of Right to the City,[16] citizens are experimenting with forms of culture that clearly depart from the hegemonic ones, as well as with new forms of political activism and new tactics for promoting civic engagement, such as *Occupy Wall Street*[17] or *Nuit Debout*: "Changer la ville pour changer la vie".[18] Such spaces may be viewed as heterotopias, as defined by Foucault, that is to say, places that are truly alternative to all the places to which they refer and of which they speak. While utopias "are arrangements which have no real space" (Foucault, 1997, p. 332) "the heterotopia has the power of juxtaposing in a single real place different spaces and locations that are incompatible with each other" (ibid., p. 335)

In conclusion, let us return to the world of literature, to recall the closing dialogue in *Invisible Cities*, between Marco Polo and the Great Khan, engaged in an endless quest for the "perfect city", to be assembled one piece at a time, composed of fragments, instants, and open-ended signals. In response to the Great Khan, who is downcast in the face of cities and cityscapes that are often harrowing for those who frequent them, "that menace in nightmares and maledictions", and can foresee no other final outcome than the "infernal city", Marco Polo has this to offer: "The inferno of the living is not something that will be; if there is one, it is what is already here." It follows, Polo argues, that there are two possible strategies for making city life more bearable: accepting the inferno and integrating into it, or the more demanding route of searching for places and people that are not part of the inferno and striving to allow them space.

Notes

1 Part of this chapter, here presented in revised and extended form, was previously published in Schmidt di Friedberg (2008).

2 The ease with which one can get lost and fail to recognize landmarks in Tokyo remains a stereotype to this day: the two main characters in the film *Erleuchtung Garantiert* (*Enlightenment Guaranteed*), directed by Doris Dörrie in 2000, get lost in the city on the very evening they arrive and end up pitching a tent for themselves in the midst of the skyscrapers.

3 Between 1955 and 2000, the number of the city's inhabitants grew by 20 million. Statistical data forecast now a population decrease (Japan Census, 2015).

4 Tokyo's rhizomatic labyrinthine structure and how it may develop in the future is taken up by Ohno (2006, p. 69).

5 The enquiry into sexual orientation (see Chapter 4) has called into question and deconstructed the identity of the flâneur, traditionally assumed to be a man: "Is the *flâneur* a transvestite? Can s/he be a cross-dressed lesbian? It is possible that the *flâneur* is a borderline case […] an example of potential, indeterminate sexuality, trapped in transliteration, caught in desire" (Munt, 1995, p. 117).

6 Henri Lefebvre met Guy Debord in 1957 and attended the Situationist International up to 1962: "It was an extremely interesting and active group, which came together in the 1950s, and one of the books that inspired the founding of the group was my book *Critique of Everyday Life*. That's why I got involved with them from such an early date" (*Henri Lefebvre on the Situationist International*, in an interview conducted and translated by Kristin Ross, printed in *October 1979*, Winter 1997, http://www.notbored.org/lefebvre-interview.html, accessed 9 April 2017). After reading Lefebvre's *La Somme et le Reste*, Guy Debord wrote to André Frankin in 1960: "The situation, as a created, organized moment […] includes perishable instants, ephemeral and unique. It is a totalizing organization that controls and favors such chance instants. From the perspective of the Lefebvrian moment, the constructed situation, then, is pitted against the instant, but at an *intermediary stage*

between instant and moment" (from *The Theory of Moments and the Construction of Situations, Internationale Situationniste* #4 (June 1960), translated by Paul Hammond, http://www.cddc.vt.edu/sionline/si/moments.html, accessed 9 April 2017). Debord and Lefebvre's relationship was intense and complex, with Lefebvre later describing it as "a sort of unfinished love affair" (Lefebvre, 2004, p. 13), and Merrifield labelling it a "Faustian fusion"; the two intellectuals broke off relations in 1963 on the head of strong theoretical divergences and an accusation of plagiarism.

7 In relation to the New Babylon, Philippe Simay has observed that: "The advent of homo ludens will turn urban space into a gigantic playing field, offering opportunities for ever new encounters, experiences and surprises. This constant renewal will be ensured by a dispositive of systematic disorder and disorientation. There, once again, we will find the model of the labyrinth" (Simay, 2008, p. 211).

8 For Henri Lefebvre, the city is a "concrete utopia" (Lefebvre, 1970), "a collective work" (Lefebvre, 1968) to be constructed via "the collective ownership and management of space" (Lefebvre, 2000 [1973]).

9 Today's city contrasts with the slower rhythm of the village or of the traditional city quarter, see Pratolini (1943).

10 Cfr. Marcuse (1985).

11 www.geofilosofia.it/metropolis/Resta_stranieri.html. Accessed 9 April 2017.

12 In Thomas Eloy Martinez's novel *The Tango Singer*, the main character wanders about the streets of labyrinthine Buenos Aires in the hopes of hearing the mysterious Julio Martel sing, and searching for Borges' Aleph. For a commentary on Eloy Martinez's work, see Masotti (2012).

13 See Queneau (1959).

14 See also the collection of disorientation guides from colleges across North America.

15 Cresswell (1996).

16 The right to the city "cannot be conceived of as a simple visiting right or as a return to traditional cities. It can only be formulated as a transformed and renewed right to urban life as long as the 'urban' [is a] place of encounter, priority of use value" (Lefebvre, 2004, p. 158).

17 "Spreading from city to city, the tactics of Occupy Wall Street are to take a central public space, a park or a square, close to where many of the levers of power are centered, and, by putting human bodies in that place, to convert public space into a political commons – a place for open discussion and debate over what that power is doing and how best to oppose its reach. This tactic, most conspicuously re-animated in the noble and ongoing struggles centered on Tahrir Square in Cairo, has spread across the world [...] It shows us that the collective power of bodies in public space is still the most effective instrument of opposition when all other means of access are blocked" (Harvey, 2012, pp. 161–162).

18 "Changing the city to change life".

9 Lost in an unfrequented wilderness[1]

Seas and forests: unjelled, wild, and threatening manifestations of nature

The theme of this chapter is disorientation in the wild, and more specifically in the "seas and forests" that Tuan has referred to as "unjelled, wild, and threatening manifestations of nature" (Tuan, 2001, p. 2). The open sea and the forest offer narratives of disorientation, of unknown islands and sailing off course, and of "another" world at the limits of human reason:

> In all the lore of woods and wilderness, no story is quite so dramatic as that of the lost person. Man's struggle with a hostile environment, the threat of death from exposure or starvation, the separation from loved ones: these and other themes are all embodied in the tale of the lost prospector, the downed pilot, or the child who has strayed from camp. Indeed, the literature of the world abounds with fictional characters who lose their way in forest, desert, cave, or sea, isolated from humanity, wandering blindly toward tragedy or adventure, from the myths of Homer and the fairy tales of Hans Christian Andersen, to the novels of Defoe and Twain. Every traveller knows them.
>
> (Hill, 1998, p. 1)

Among the various "wild, and threatening manifestations of nature", I have chosen to focus on these two in particular by virtue of their rich symbolic and imaginary value: the forest, "a trackless region of possibility" (Tuan, 2007, p. 56) and the sea: "our endless saga, the sea is our most powerful metaphor, the ocean in us" (Epeli Hau'ofa, cit. in Mack, 2011, p. 73). However, I might have chosen other physical places that suggest a similar wealth of interpretations and have also inspired a body of literary descriptions: the desert, for example, the mountains, caves,[2] snow-covered landscapes,[3] fog, or the deep waters explored by scuba divers: "a one-dimensional space. One substance, one dimension. And we are so remote from the earth and life on earth, that this dimension of water bears the mark of limitlessness. To try and find high, low, right or left in a world that is so well unified by its substance, […] is not living in the new world conquered by diving" (Bachelard, 1957, p. 186). In the wild, terrain and climate conditions combine to pose a particular challenge to human orientation, and cultural systems have grappled with this challenge since ancient times,[4] seeking ways of adapting to the environment and interpreting the landscape; nonetheless, our perceptions of landscape remain predominantly subjective: "Strange to think that the question 'Who am I?' can be answered by a landscape" (Tuan, 2001). Yi-Fu Tuan has described his "affinity" for desert landscapes in particular, an attraction that he experienced as love at first sight and attributes to his great dread of getting lost:

> Disorientation is a specific fear that [...] is far less a pall on life, incapacitating one for bold action, than diffuse and persistent anxiety. [...] if any place reassures me – restores my mental balance – it is the sterile desert. [...] I have come to see that the desert's profound appeal to me lies not only in its pure lines and ease of orientation but in barrenness itself – in the presentation of absence that enables me to wipe out in one clean sweep sex, biological life, and death.
>
> (ibid.)

Disorientation, Tuan believes, is also implicated in his deep antipathy to forest environments: "I dislike the tropical rainforest for the same reason: unless one is a native, it invites disorientation. I take to the desert because it is an open map, with the sun serving as a dependable marker of east and west, and with sharply-etched landforms – visible from miles away – that unmistakably tell visitors where they are" (Tuan, 2009, p. 64). Our interpretation of the physical environment, the topophilia described by Tuan, is inevitably informed by sociopsychological categories:

> The rainforest has the opposite effect: by forcing me to confront choking growths and struggle, it rubs my nose in the pungent odor of decay, which fuses in my mind with the pungent odor of sex. [...] But I have another reason for my dislike of the rainforest – its threat to individuality. Death, of course, means the end of individuality and reabsorption into an undifferentiated whole. But the rainforest denies individuality in its very superfluity of life. In the dense biomass, no single plant, animal, or human being can stand out.
>
> (Tuan, 2001)

While in general "landscapes refuse to be disciplined [..., making] a mockery of the oppositions that we create between time and space, or between nature and culture" (Massey, 2005, p. 136), getting lost in the landscape of the wilderness, with its elusiveness, ambiguity, and rich symbolism, is the story of our relationship with nature in the wild state. With its explosive materiality and diversity, the wild is the setting in which the encounter between dreams and reality, nature and human desire, order and disorder invariably plays out:

"The wild occupies a special place in the imagined empires of human civilization as that which lies outside its historical and geographical reach, however defined"[5] (Whatmore, 2002, p. 12). The mysterious life of the non-human, domain of all that is wild and irrational, stands in contrast with the rational order of human thought and the planned city. In his *Discourse on the Method* (1637), Descartes avows that he will be firm and resolute in applying his method and avoid being swayed by unsubstantiated opinion:

> Imitating in this the example of travelers who, when they have lost their way in a forest, ought not to wander from side to side, far less remain in one place, but proceed constantly towards the same side in as straight a line as possible, without changing their direction for slight reasons, although perhaps it might be chance alone which at first determined the selection; for in this way, if they do not exactly reach the point they desire, they will come at least in the end to some place that will probably be preferable to the middle of a forest.[6]
>
> (Descartes, 2009, pp. 34–35)

The green living world of the woods is the very opposite of the inhabited and constructed world of humans; it is the emblematic place of our encounter with the non-human, the unexpected, as we are reminded by each small sound or movement that we perceive there: "Alone in the forest in the solitude that is part night, part dawn, part trees, part animals, and that gives one the momentary sensation that one has lost one's way in the world and must some day retrace one's steps to this wild and dangerous place that is truly home" (Màrai, 1998, p. 118).

"I found that I was in a gloomy wood, because the path which led aright was lost"

> "Nel mezzo del cammin di nostra vita
> Mi ritrovai per una selva oscura
> Chè la diritta via era smarrita
> Ahi quanto a dir quale era è cosa dura
> Esta selva selvaggia e aspra e forte
> Che nel pensier rinnova la paura!»
> "When half way through the journey of our life
> I found that I was in a gloomy wood,
> because the path which led aright was lost.
> And ah, how hard it is to say just what
> this wild and rough and stubborn woodland was,
> the very thought of which renews my fear!"
>
> (Dante Alighieri, *Inferno*, Canto I, vv. 1–6)[7]

Throughout the literary tradition, the woods recur as a metaphor of disorientation in both the moral and the physical senses. In the *Divine Comedy*, getting lost in the woods is an allegory for Christian guilt and straying from God: "Bestiality, depravation, sin, perdition – this is what forests were associated with in Christian mythology" (Harrison, 1992, p. 76). The "gloomy wood" is impenetrable to divine light, the soul strays from the path of righteousness and loses itself in sin. However, the woods provide a backdrop to all kinds of human behaviour: in Dante's *Purgatory* we find another wood, the "divine forest", which is brighter and more open. Dante makes his way into this unknown wood – "My slow steps had carried me so far inside the ancient wood, that I no longer could see whence I had entered it" (Dante, *Purgatorio*, Canto XXVIII, vv. 22–24, 1918, p. 47) – but does not get lost, because he is now on his way to redemption; here he will meet Matelda, the personification of perfect happiness before original sin. Other medieval authors used the woods as a moral metaphor: some decades before Dante, the Italian philosopher and rhetorician, Brunetto Latini, was returning from a diplomatic mission to the court of Alphonso X in Castile, when he learnt that the Florentine Guelphs had been defeated by the Sienese Ghibellines in the Battle of Montaperti (1260), in the context of the Guelph-Ghibelline war then underway in Tuscany. Greatly distressed, the learned scholar, whom Dante considered his teacher,[8] wandered sadly on and lost his way: "And I, in such anguish, thinking with head downcast, lost the great highway, and took the crossroad through a strange wood" [E io, in tal corrotto pensando a capo chino, perdei il gran cammino, e tenni a la traversa d'una selva diversa] (vv. 185–190).[9] In this "strange wood", he went on to meet fantastic creatures, plants, and animals (and even the geographer Ptolemy). Thus, in Latini, as in Dante, getting lost took on a metaphorical meaning, becoming a journey in search of salvation and truth. But, why does the forest – or the woods – so consistently feature as a place of disorientation?

In the Middle Ages, forests played a key role in the daily life of people all over Europe. The extensive forest cover represented a crucial economic resource (providing wood for building and energy, in addition to game, fruits, honey, grazing for animals, etc.), but also had major implications for transport and mobility: the woods could not be ignored at either the physical or the imaginary level: "Landscapes are culture before they are nature: constructs of the imagination projected onto wood and water and rock. [...] But it should also be acknowledged that once a certain idea of landscape, a myth, a vision, establishes itself in an actual place, it has a peculiar way of muddling categories, of making metaphors more real than their referents: of becoming, in fact, part of the scenery" (Schama, 1995, p. 61). The medieval forest, by virtue of its considerable size, was an inevitable part of the population's everyday lives, intensively exploited, and an ever-present and indispensable landmark.

Nonetheless, forest cover in Europe has fluctuated over the past 1,500 years: a substantial body of research (see for example Bork et al., 1998), based on pollen analysis and the study of settlements, has informed scholarly reconstructions of variations in land use from the Neolithic Era onwards. The space occupied by woodlands and that taken up by agricultural land have historically competed with one another within a dynamic system influenced by political, demographic, economic, technological, and infrastructural change. If we take the situation in Germany (Table 9.1) as an example, it appears that forest cover peaked in the seventh century (90 per cent), later decreasing sharply in the early 1300s (15 per cent) due to the expansion of built-up areas and cultivated land: at that time, the reclaiming of swamplands by monastic orders and the rotation of crops had contributed to demographic growth and a rise in the domestic animal population. The early 1400s saw a strong recovery in the proportion of wooded areas (45 per cent), in the wake of the Black Plague; the forests subsequently shrank once more, stabilizing at around the 30 per cent mark (Bork et al., 1998). The situation in Southern Europe was not drastically different, suffice it to recall the unprecedentedly aggressive deforestation policy implemented by the Republic of Venice between the fourteenth and sixteenth centuries. In Dante's time, therefore, the forest cover had reached its historic low point. Furthermore, recent data (Lavalle and Silva, 2011), indicates that in Europe today the level of forest cover – though altered due to human intervention, with a higher proportion of conifers and a lower proportion of deciduous varieties – is still considerable. The woods continue to be part of the landscape that surrounds us, although we have distanced ourselves from them and from the wildlife that inhabits them; we are not lacking in forest, it is just that we no longer have an intimate everyday relationship with it.

By barely one generation after Dante, getting lost in the woods had taken on a more worldly and secular value. For the civic commune of Boccaccio's (1313 –1375) Florence, the forest was still a place where one went astray, though no longer in terms of losing the divine light, but rather in the sense of undertaking an initiatory rite: "If the forest represented a dark and gloomy

Table 9.1 Land use in Germany, excluding the Alps, within current borders (the figures are expressed in percentage terms, the remaining area is occupied by water, infrastructure, and public parklands).

Year	*Forests*	*Cultivated Land and Pasture*	*Period*	*Forest*	*Cultivated Land and Pasture*
650–659	90	5	1420–1429	45	28
750–759	87	7	1520–1529	34	38
900–909	68	18	1608–1617	30	41
1000–1009	65	20	1650–1659	32	32
1250–1259	20	51	1780–1789	30	39
1310–1319	15	55	1870–1879	27	40
1340–1349	17	54	1961–1990	30	38

Source: Bork et al., 1998, p. 161, adapted

wood, the Church did all in its power to prevent it being illuminated, and from the High Medieval Period onwards, unleashed all its hostility against it. The hagiographers of the day described the woods, in the best case scenario, as a stronghold of pagan worship, and in the worst, as the scene of orgies, home to witches and to every form of bestiality" (Paci, 2002, pp. 31–32). In the *Story of Pietro and Agnolella* in the *Decameron* (1349–51), a young couple elope but become separated from one another during their flight, and end up getting lost in the woods:

> Pietro [...] getting on horseback rode as fast as he could [...] and went up and down the forest, calling aloud to see if she could hear, but in vain [...] In this manner did poor Pietro traverse the forest over and over, hallooing and shouting, and frequently coming back again, when he thought he was going forward all the time. [...] The young lady, in the meantime, [...] was carried out so far into the wood that she could not find the way out again: therefore, she went up and down full of grief for what had happened.
>
> (Boccaccio, 1881, *The Fifth Day*, Novel III, pp. 240–241).

Experiencing the violence and animality of the woods becomes an initiatory journey towards married life and sexuality for the two young people, who are subsequently able to marry in the safety of the castle. Jones and Cloke have argued that "there is a clear gendered dimension to woods as landscape of fear" (Jones, Cloke, 2002, p. 28). "The 'forests' are thereby imbued with notions of danger, mystery and lack of light, and with a clear sense of the risk to women and children of becoming lost, far from home. [...] Forests are thus likely to be important as areas of women's fear" (ibid., pp. 24–25). Hence the misadventures of Tom Thumb, Hansel and Gretel, Little Red Riding Hood, Snow White, and numerous other fairy-tale characters whose stories run on the theme of getting lost. Children that are abandoned in the forest must face up to its dangers and overcome their own fears: on reaching adolescence, they are forced to leave home and to undertake the journey towards adulthood and maturity.

The forest is the place of danger, a form of moral disorientation. The knights in the Arthurian literary cycle, Lancelot, Ywain, Tristan, all get lost in the forest of Broceliande. The characters in Ludovico Ariosto's epic poem *Orlando Furioso* (1532) encounter but then lose one another in the thick forest, scattering in different directions. While Angelica gets physically lost – "Through dreary woods and dark the damsel fled / By rude unharboured heath and savage height / While every leaf or spray that rustled, bred / (Of oak, or elm, or beach), such new affright / She here and there her foaming palfrey sped / By strange and crooked paths with furious flight" (Ariosto, I, 32–35, p. 14) – Orlando actually loses his reason. For the "knights errant" getting lost in the woods was an "error": forests fell outside of the established order and the rules of urban life. Losing oneself in them was therefore a rite of passage preparing one to subsequently obey the rational and legitimate behavioural codes of the city (Harrison, 1992). The forest brings one "into contact with the wonderful and the adventurous, a kind of labyrinth in which one can get lost or find oneself, win or be defeated, experience pleasant emotions or go mad. The forest, site par excellence of uncertainty and indeterminateness, is the funnel through which to channel the tensions of eros, separating the search for what is good from the temptation to sin" (Paci, 2002, p. 38). After their temporary madness, both Orlando and Ywain return to normal and join a higher social order in which there is no room for error or derangement. The same pattern applies to forest outlaws Robin Hood and Gamelyn, who find refuge from oppression in the woods but return to the established order and the city once they have successfully made justice prevail (Harrison, 1992). Furthermore, forests are inhabited by: "Crowds of strange beings endowed with superhuman powers and characters, although partaking human form" (Porteous, 2002, p. 84); they provide a theatre for chaos, disguise, and the undertakings of witches, elves, druids, and pagan divinities. In *A Midsummer Night's Dream*, Lysander – gone astray in the enchanted wood – declares: "Fair love, you faint with wandering in the wood; And to speak troth, I have forgot our way" (Act 2, Scene 2, 23) while Queen Titania warns the donkey-headed Bottom: "Out of this wood do not desire to go: Thou shalt remain here, whether thou wilt or no. I am a spirit of no common rate" (Shakespeare, Act 1, Scene 1, 75).

Holzwege

The word forest derives from the Latin *foris*, meaning outside: it was originally a legal term used in Lombard edicts to refer to woods outside the jurisdiction of the town, which were to be exclusively used by the king as hunting grounds (Harrison, 1992). The terms *wald, wood*[10] on the other hand, may be related to the expression *wild* (Konijnendijk, 2008). For Dante, getting lost in the "wild and rough and stubborn woodland" was a terrifying experience, "the very thought of which renews my fear!" (Dante, *Inferno*, Canto I, vv. 5–6, 2003, p. 1). Experiencing fear conditions how we perceive places: "Fear is an important part of landscape and of the tension between place (home, safety) and space (the unknown, the adventurous and maybe unsafe)" (Konijnendijk, 2008, p. 35). In the German-speaking world, the term *Waldangst* has been coined to signify the sense of fear,[11] often irrational, elicited in city-dwellers by the woods; those who live in the woods and off the woods, in contrast, have become familiar with the woodland life and environment, and have therefore learnt not to be afraid of it: "Forest people don't get lost!" (Jonnson, 2002, p. 96). Jones and Cloke (2002) distinguish between two categories of fear in relation to the forest, the fear of what the woods may contain (dangerous animals, poisonous plants, lightening, fires, spirits, elves, witches, wild men, bandits, criminals) and a deeper fear, that of getting lost:

> The forest, no less than the bare plain, is a trackless region of possibility. Trees that clutter up space from one viewpoint are, from another, the means by which a special awareness of space is created, for the trees stand one behind the other as far as the eyes can see, and they encourage the mind to extrapolate infinity. [...] The forest although it may be small, appears boundless to one lost in its midst.
>
> (Tuan, 2007, p. 56)

The fear of becoming lost in the woods is the fear of unknown space, of that which is different, of being immersed in a place that is dominated by other sensory categories, by sounds, noises, smells, and movements that are unusual for those who come from the city.[12] In the confined space of the forest: "Humans are enclosed, dwarfed, concealed and/or lost" (Konijnendijk, 2008, p. 15).

The myth of the *deutsche Wald*, "the German woods", is a product of the Romantic Era that was represented through painting (Caspar David Friedrich, *Der Chasseur im Walde*, 1814), music, and poetry: the forest transmits a sense of melancholy, fear, and mystery, as portrayed in the tales of the Brothers Grimm, because, as Lehmann has argued, "for many of us, the inside of the woods remains something 'disturbing' and 'full of secrets'. In any case it is somewhere that we can easily get lost" (Lehmann, 1999, p. 8). In the labyrinthine descriptions of Kafka, disorientation takes on a psychological and introspective dimension, a wandering about one's own being rather than the green and natural surroundings of the woods (see *The Burrow*). Nonetheless, in his bibliographical notes, the writer used the forest to describe his own sensation of being lost:

> In the forest, I was completely lost. Incomprehensibly lost, because up to a short time beforehand even though I hadn't been following a path, I had been walking close to and in full view of one. Now I was lost, the path had disappeared, all my attempts to find it had failed. I sat down on a tree trunk to review the situation, but I was distracted, my thoughts were continuously straying and not necessarily to more important matters, I was relegating my worries to my dreams. Then I noticed that some of the plants around me were covered in berries, I gathered some and I ate them.
>
> (Kafka, 2016, p. 1517)

Terror and mystery, especially at night, are part of the aesthetic experience of the German woods, as echoed in the dismal cry of the witch Lorelei: "Es ist schon spät, es wird schon kalt. Kommst nimmermehr aus diesem Wald"[13] (von Eichendorff, 1984, p. 306). The woods contributed to the construction of German identity and naturalism, in that they represented nature, native purity, courage, and tradition. German language and culture are rooted in the legends and fairy tales of the Middle Ages; there appears to be an affinity between the wood and the German personality that stretches back to the Germany described by Tacitus with its primeval forests; and to the mythical enterprises of Arminius-Hermann in the forest of Teutoburg, narrated by Klopstock (1760). The Romantic construction of the *deutsche Wald* would later degenerate into the notions of *Heimat*, and *Naturschutz*, with their overtones of racism and exclusion, ultimately becoming a vehicle for Nazi propaganda (an example of this is the 1936 film *Dauerwald*). Anna Bramwell observed that Nazi Germany was the first European nation to introduce detailed legislation for protecting nature; in 1934 specific provisions for reforestation with native tree species were introduced[14] along with a landscape protection policy aimed at providing a "vital space for the German people".[15] The reforestation programme became loaded with symbolic value related, for example, to the oak tree, whose leaf was a key symbol for the SS (Bramwell, 1989): these developments were underpinned by an exaggerated ideal of purity that drew on a primordial archetype.

In contemporary cities, the woods and woodland life, though still present, are no longer part of our daily lives, no longer part of ourselves, similarly to other wild places: "'Territorial anguish' prefigures the loss of a place's belonging to a human group and vice versa. [...] Today there is a form of getting lost that consists of a permanent sense of distraction vis-à-vis one's surroundings" (La Cecla, 2000, p. 37). Getting lost in the woods therefore implies alienation. We find ourselves walking along Heidegger's *Holzwege:* "Wandering is seen by Heidegger as the destiny of the West and perhaps that of the human being in general, for 'the soul is a stranger upon the earth'" (Resta, 1996, p. 13). In his contemplation of the landscape and woods of Todtnauberg, Heidegger was not in search of a nostalgic return to the past, in the sense of *Blut und Boden* [*Blood and Soil*], but condemning the loss of a sense of place and of nature, and the "rootlessness" that is now an intrinsic part of our way of life. In Heidegger's view, the human being of the late modern era, the "atomic era" of the "consumer civilization", has become "a functionary of technology" (*Angestellter der Technik*) (Heidegger, 2002, p. 220), losing the capacity to inhabit places: all that remains are "cut-off paths" (*Holzwege*) and *Lichtungen*, "forest clearings hidden by trees": "Forest clearings form a centre that is not always possible to enter; from the edge one looks in and the appearance of animal tracks does not encourage one to advance any further" (Zambrano, 1988, p. 11). *Holz*, Heidegger explained: "is an old name for forest. In the wood [*Holz*] there are paths [*Wege*], mostly overgrown, that come to an abrupt stop where the wood is untrodden. They are called *Holzwege*. Each goes its separate way, though in the same forest. It often appears as if one is identical to another. But it only appears so. Woodcutters and forest keepers know these paths. They know that it means to be on a *Holzweg* [*auf einem Holzweg zu sein*]" (Heidegger, 2002, p. V).

Over-forestation

Although at the global level, deforestation continues apace,[16] many countries maintain a significant level of forest cover – mostly accounted for by non-native varieties – due to the shift away from agriculture, the spread of industrial plantations, and reforestation policies. In Japan,[17] despite the urban explosion of recent decades, two-thirds of the territory are still covered by mountain forests (*yama*),[18] leading the country to be dubbed "the green archipelago" (Totman, 1989). The original vegetation cover comprised beech woods (*bunarin*) at higher altitudes (700–1500 m) and in the north-east of the country, and warm-temperate forests of lucidophyllus ("shiny leaf") species such as the evergreen oak, Pasania and Camellia (*shoyojurin*) in the south-west and low-lying areas (below 500 m), and provided the population with food, wood, and firewood. Throughout Japanese history, periods of intense deforestation have cyclically alternated with periods of reforestation: the former trend peaked in the period running up to and during the Second World War, at which time most of the original forest cover was destroyed, with serious consequences at the hydrogeological level and economic repercussions due to the resulting scarcity of wood. Reforestation was particularly vigorous in the post-WWII period, with the outcome of restoring the mountain forest cover, but also wiping out most of the original varieties; nowadays about half of the Japanese forests have been replaced by industrial plantations of conifers, particularly cypresses (*hinoki, Chamaecyparis obtusa*) and cryptomerias (*sugi, Cryptomeria japonica*), while the original tree species account for no more than 5 per cent of the total (Knight, 2000).

In Knight's view, Japan currently faces the challenge of over-forestation in the interior of the country,[19] where forests have often fallen into a state of abandon. In many mountain areas, the villages are under siege from the surrounding vegetation, leading in some cases to their abandonment; the conifer plantations, no longer cut back or exploited for commercial purposes, make the mountainsides more vulnerable to landslides, while wild animals (such as monkeys, deer, and wild boar) invade the built-up areas: "In addition to the unwanted embrace of the new forest, animal invasions signal that the village is no longer fit for human habitation [...]. Ultimately, the aggressive forest is a mirror which reflects the inability of the present generation of villagers to control it as past generations have done" (Knight, 1997, pp. 721–722). Thus, the forest with its power to symbolize origin and rebirth, but also a place of terror, continues to occupy a prominent place in the Japanese imaginary and representations. This is reflected in *manga* and *anime* (see Hiao Miyazakie's *Princess Mononoke*) as well as in contemporary literature, from Abe Kobo to Murakami Haruki: "Now I know exactly how dangerous the forest can be. [...] I'd never imagined that trees could be so weird and unearthly. [...] They have a physical power" (Murakami, 2005, pp. 141–142). Murakami continues: "I'm totally lost, my identity dying. There's no direction where I am, no sky, no ground. [...] So there is nothing left to fear" (ibid., 2005, pp. 405–406).

The forest of Aokigahara stretches over an area of 3,000 hectares on the slopes of Mount Fuji. This "sea of trees", traditionally associated with the presence of the Yūrei, the ghosts of Japanese folklore, is sadly famous today as a popular site for suicide.[20] The rich and dense vegetation of its volcanic caves block GPS or mobile phone coverage, and walkers and mountain bikers mark their paths with lengths of coloured sticky tape so as not to lose their way. This practice is harmful to the trees and the forest rangers are engaged in a constant battle to remove the tape.

"Lost in an unfrequented wilderness"

The wilderness and the image of the forest are also recurrent themes in American Romanticism and neo-Gothicism: they feature in the verses of Emerson, and the paintings of Cole, Moran, and Bierstadt, as well as in the novels of Fenimore Cooper and Nathaniel Hawthorn: "The path strangled onward into the mystery of the primeval forest" (Hawthorn, *The Scarlet Letter*, 1850). Aesthetic admiration of the forest was bound up with the awareness that it was rapidly disappearing: in two long poems, *The Complaint of the Forest* (1835) and *The Lament of the Forest* (1841), the painter Thomas Cole mourned the forest, under assault from the woodcutter's axe. "Progress" seemed inevitably bound to destroy the cultural and natural world encountered by the first Europeans, leaving nostalgia for a primitive charm that was being cancelled out by modernity. Under the influence of Trascendentalism, the notion of preservation took on religious and spiritual overtones; for Emerson, the woods were a holy temple to which human beings went to purify themselves and nature was the word of God: "In the woods, we return to reason and faith" (Emerson, 1983, p. 10). Thoreau celebrated the *wilderness* for its regenerative power: "I went to the woods because I wished to live deliberately, to front only the essential facts of life, and see if I could not learn what it had to teach, and not, when I came to die, discover that I had not lived" (Thoreau, 1990, p. 60). Hence, getting lost in the woods bore the meaning of a return to the primeval state, to freedom in the midst of nature:

> It is a surprising and memorable, as well as valuable experience, to be lost in the woods any time. […] In our most trivial walks, we are constantly, though unconsciously, steering like pilots by certain well-known beacons and headlands, and if we go beyond our usual course we still carry in our minds the bearing of some neighboring cape; and not till we are completely lost, or turned round, […] do we appreciate the vastness and strangeness of Nature. Every man has to learn the points of compass again as often as he awakes, whether from sleep or any abstraction. Not till we are lost, in other words, not till we have lost the world, do we begin to find ourselves, and realize where we are and the infinite extent of our relations.
> (ibid., p. 105)

In the cult of the *wilderness*, getting lost is a terrifying, yet sublime experience. A case in point is the dramatic narrative *Thirty-Seven Days of Peril* by Truman C. Everts, a member of the 1870 Washburn-Langford-Doane Expedition to unexplored areas of Yellowstone. Everts got cut off from the rest of the party and found himself lost within a vast expanse of forest; he dismounted to search for landmarks and his horse ran away with his entire store of provisions and equipment. During his 37 days in the heart of Yellowstone, Everts was: "lost in it, without any of the ordinary means of subsistence, and [...] wandering for days and weeks, in a famishing condition, alone, in an unfrequented wilderness" (Everts, 1871, p. 1). Initially Everts feared an encounter with "Indians", while subsequently his greatest desire became that of meeting them; throughout all his wanderings, however, he never encountered a single human being. As the days went by, Everts' situation became ever more desperate, and his fear of the wilderness escalated: "I peered upwards through the darkness, but all was blackness and gloom. The forest seems alive with the screeching of night birds, the angry barking of coyote, and the prolonged, dismal howl of the gray wolf", but he bravely resolved: "not to perish in that wilderness" (ibid., p. 3). Though starving and wounded, he persevered in searching for a way out, despite nights of terror, cold, and attacks by wild animals.

Everts' detailed account of his days of terror allows us to identify another characteristic of the ideal of "white supremacy": an aesthetic appreciation of the wilderness that goes hand in hand with the ability to engage in abstract reasoning, and in turn, is clearly linked to European colonialism (Schmidt di Friedberg, 2004). The North American concept of wilderness, understood as virgin nature without any trace of occupation by humans, implicitly "dehumanizes" the indigenous populations who live or have lived in contact with it, justifying expropriation by the colonial powers. Callicot and Nelson observe that the notion of wilderness is "alleged to be ethnocentric, androcentric, phallogocentric, unscientific, unphilosophic, impolitic, outmoded, even genocidal" (Callicott and Nelson, 1998, p. 2). In the midst of the most extreme situations and moments of greatest hardship, Everts admired lakes, mountains, and animals in their extraordinary natural setting and longed to share his aesthetic enjoyment with family and friends: "The grandest landscape I ever beheld. [...] this display of secluded magnificence, now, probably for the first time beheld by mortal eyes" (Everts, 1871, p. 3).

The height of Everts' drama occurred one night when he had succeeded in starting a fire, accidentally setting the forest alight. Although frightened and in pain from burns, even on this occasion, Everts managed to admire the spectacular sight:

> The grandeur of the burning forest surpasses description. An immense sheet of flame, following to their tops the lofty trees of an almost impenetrable pine forest, leaping madly from top to top, and sending thousands of forked tongues a hundred feet or more athwart the midnight darkness [...] I never before saw anything so terribly beautiful. [...] The roaring, cracking, crashing, and snapping of falling limbs and burning foliage was deafening. On, on, on travelled the destructive element, until it seemed as if the whole forest was enveloped in flame.
>
> (ibid., p. 10)

However, as the days went by and Everts' physical condition deteriorated further, even the landscape lost its appeal: "The great falls of the Yellowstone were roaring within three hundred yards, and the awful canon yawned at my feet; but they had lost their charm for me. In fact, I regarded them as enemies which had lured me to destruction, and felt a sullen satisfaction in morbid indifference" (ibid., pp. 12–13).

John and Metzo have revisited Everts' adventure, exploring the impact of his dramatic experience on the visibility and fame of Yellowstone Park. Following the phenomenological approach of Merleau-Ponty and non-representational theory, these authors observe the role of the body and its symbolism in the practice of exploration and the complex relations between body and landscape, identifying two key underlying themes: "The emergent ideology of a national (and ostensibly white) manhood [...] and [...] the Romantic appreciation for the sublime in viewing landscape [...] each drawing on the dominant Cartesian and Romantic dualism of mind/matter, man/nature and body/landscape" (John and Metzo, 2008, p. 231). Everts' 37-day ordeal became "the single-most sensational embodied explorative encounter with the region" (ibid, p. 226). Everts, male and white, does not appear to have had a particularly strong physique, but by virtue of his strength of character he "personified the ideal of American manhood" (ibid., p. 230). His determination "not to perish in that wilderness" (Everts, 1871, p. 3) became: "in Cartesian and moral terms [...] a physical and psychical contest between self and world" (John and Metzo, 2008, p. 229).

All at sea

> Ignoranti quem portum petat nullus suus ventus est.[21]
>
> (Seneca)

The entire history of geographical "explorations" and the human colonization of the earth is made up of extraordinary endeavours, whose protagonists set out "not to refuse experience of that world behind the sun which knows not man!" [non vogliate negar l'esperienza, di retro al sol, del mondo sanza gente!] (Dante, Inferno, Canto XXVI, 116–17). No less rich, but certainly less documented, is the history of loss of orientation during sea and land voyages. The sea in particular has tragic stories to tell of shipwreck and voyages gone astray, but also of chance encounters, mysterious islands, and great discoveries that came about thanks to episodes of disorientation: "In English the phrase 'all at sea' carries with it the sense that someone is completely and utterly lost: This is the sea as wilderness, as a place without paths or distinctive marks. Once in its midst, the fear of becoming irretrievably disoriented is inescapable" (Mack, 2011, p. 72).

In the Odyssey, Homer narrates Ulysses' epic voyage through the Mediterranean prior to 1000 BC: almost 10 years and 10,000 kilometres of wandering were required to cover the 500 kilometres from Troy to Ithaca. On several occasions, such as at Cape Maleas, the southernmost point of Greece, Ulysses notes that he is lost: "But as I rounded Malea a rushing wave and a North Wind pushed me off and drove me past Cythera" (Odyssey, IX, 80–81). Christopher Columbus' famous "error" and his arrival in 1492 at the "islands of India beyond the Ganges" represent another geographical paradox. The plan to *buscar el levante por el poniente* (seek the east via the west) by sailing westwards towards India, thereby avoiding the long and dangerous circumnavigation of Africa, appears to have been based on a dual error of calculation, an underestimate of the true circumference of the earth, and the incorrect conversion of Arabic miles into nautical miles. Allegedly based on mistaken input from the Florentine scientist, Paolo dal Pozzo Toscanelli,[22] Columbus estimated the distance to be travelled at one-third of its true extent, or 6,500 miles rather than 20,000, a long crossing but feasible with the resources of the day (http://www.arcetri.astro.it/CDAs). Andrew Sharp has claimed that the "discovery" of Greenland by the Vikings, like that of America, was accidental and not the outcome of planned voyages (Sharp, 1957). The theory of the accidental discovery of Greenland is also taken up by Huth who attributes it to an error of navigation:

> The first sighting of Greenland, somewhat before AD 1000, also appears to have been due to an accident, when Gunnbjörn Ulfsson, sailing from Norway to Iceland, was blown off course. By legend, Erik Thorvaldsson (Erik the Red) established the first colony in Greenland soon thereafter. A number of settlements along the south coast of Greenland were established but died out around the fourteenth century.
>
> (Huth, 2013, p. 4)

The human relationship with the sea varies widely across cultures in line with political, socioeconomic, and religious frameworks; thus marine space has come to bear a host of symbolic and mythological meanings and values. According to Mack:

> There are two alternative cultural views of the sea: one which constructs it as an unwelcoming wilderness where the land is a reassuring point of reference; the other which sees it as entirely familiar and unthreatening. Those are, however, only characterizations of two extremes. Arguably there are not completely maritime cultures any more than there are completely terrestrial ones.
>
> (Mack, 2011, p. 74)

Visions of the sea as a place of disorientation and terror are not necessarily linked to distance from the coast and thus irrelevant to island countries. For example, Mack has argued that the Japanese, towards the close of the Tokugawa era (1603–1868), deliberately chose not to be a "marine culture". At the end of the eighteenth century, the Japanese mathematician and politician Honda

Toshiaki (1744–1821), favourable to the development of relations with the West, scathingly commented on the poor nautical ability of his fellow countrymen:

> Even when the weather improves, the crew are at a loss to tell in which direction to head, and the ship floats about helplessly. As a last resort they cut off their hair and make vows to Buddha and the gods. Then they take out pieces of paper on which have been written the names of the twelve directions, roll them up into balls, and put them into a basket with a hole in its lid .... The captain and crew, in tears, fervently call on Buddha and the gods of heaven and earth to indicate the direction. They grasp the basket in their hands and strike the lid. Then, when one of the pellets jumps out, they pick it up, their eyes blinded by tears of joy, and cry that it is the direction vouchsafed by Buddha and the gods. They then set their course by it, and go completely astray.
>
> (ibid., p. 84)

The quest for longitude

The difficulty of planning long sea voyages was even greater when it came to crossing the ocean. Shipwreck literature became a literary genre in its own right in the sixteenth and seventeenth centuries, the era of the great European explorations (Medas, 2004). The single greatest challenge for navigation was determining longitude. The discovery of a method for measuring longitude was the first of the unattainable dreams listed by Gulliver on his visit to Laputa: "I should then see the Discovery of the *Longitude*, the *perpetual Motion*, the *Universal Medicine*, and many other great Inventions brought to the utmost Perfection" (Swift, 1900, p. 57). While latitude could be accurately calculated from the height of the midday sun with the help of special astronomic tables, defining longitude required application of the dead reckoning method,[23] which involved keeping track of the distance travelled and the direction taken from a known starting position. Losing track of the ship's route or choosing to follow a line of constant latitude, a practice known as *running down a westing* meant prolonging the voyage, at times by weeks, with serious repercussions for the health of the sailors, including starvation and scurvy, as provisions ran out. During long ocean crossings, cumulative error due to imprecise measurements of position while sailing led many ships to stray off their routes, often with tragic consequences.

One of the most sensational disasters occurred off the Isles of Schilly in 1707, when 1,400 people lost their lives. During the War of the Spanish Succession, the Royal Navy sent Sir Cloudesely Shovell, commanding admiral of the British fleet, to support the forces of the Grand Alliance of England, Austria, and Holland at the Battle of Toulon. At the end of the mission in late October, the British fleet of some 20 ships set sail from Gibraltar in bad weather conditions, bound for Portsmouth. When the ships reached the

English Channel, Shovell's sailing masters, due to an error in the calculation of their position, thought that they were in sight of the island of Ushant off the coast of Brittany located at the mouth of the channel, while instead they were driven onto the cliffs of the Isles of Schilly, 35 leagues further north (Cooke, 1883). The great shock caused by this disaster and other shipwrecks led the British government to set up a *Board of Longitude* in 1714 during the reign of Queen Anne, offering a generous financial reward to anyone who might succeed in devising a method for accurately measuring longitude during ocean crossings. This was clearly stated in the "Act for providing a Public Reward for such Person or Persons as shall discover the Longitude at Sea": "The Discovery of the Longitude is of such Consequence to Great Britain, for the safety of the Navy, and Merchant Ships, as well as Improvement of Trade, that, for want thereof, many Ships have been retarded in their Voyages, and many lost" (Nicholson and Rousseau, 1968, p. 168). The issue of longitude also became a key factor in the race among the great maritime powers of Europe to colonize the globe.

Indeed, rewards for solving the question of longitude measurement to within half a degree had already been offered in Spain by Philip II and Philip III, as well as in Holland and in France from the end of the sixteenth century onwards. The French *Bureau des longitudes*, (today's *Institut de mécanique céleste et de calcul des éphémérides*), was founded by Abbé Grégoire on 25 June 1795. For three centuries, the great names of science – Galileo, Huygens, Leibnitz, Hooke, Newton, and Cassini – grappled with the issue of longitude, bringing to bear a series of different strategies: attempts were made to develop an accurate marine chronometer – an idea first proposed by Gemma Frisius in 1530 – so that the time at the meridian from which the ship had set sail could be reliably compared with local time at its current position; the lunar distances method, based on calculating the angle between the moon and other celestial bodies; and the study of the magnetic declination of the Earth. The English clock-maker John Harrison, after 20 years of experimentation, succeeded in producing an accurate marine chronometer in 1759, but the lunar distances method prevailed until the mid-1800s due to the influence of the British astronomer Nevil Maskelyne (see Juhel, 2013). As Müller-Sievers has argued, even Harrison did not identify a definitive solution to the issue of longitude measurement, but came up with a practical ad hoc method of orientation for sea voyages. Twenty years later, the still unresolved problem inspired Kant to take an interest in (dis)orientation from a philosophical perspective, both in relation to "transitive" orientation (e.g., locating a subject within an oriented framework such as a geographical map) and "intransitive" orientation (the subject's self-determination) (cfr. Müller-Sievers, 2003, pp. 13–16).

Accidental settlements in the Pacific

An interesting and controversial theme is the history of how the Pacific islands came to be colonized by their first inhabitants, whose great seafaring tradition dates back to ancient times. On these islands, generations of expert teachers and schools in the true sense of the word developed an art of navigation, in which practice was blended with magic.[24] Neither maps nor navigation instruments of any kind were used on board the boats. The islanders memorized the position of the stars, the islands, and the motion of the waves, creating "a highly sophisticated, albeit instrumentless, way to use the heavens for navigation" (Finney, 1998, p. 489). They appear to have oriented themselves via the interpretation of a complex set of signs provided by the heavens and the sea, also using: "the entire canoe as a swell gauging instrument" (ibid., p. 487), and drawing on the perceptions of the entire body. Direction was determined by interpreting the voice of the sea in all its forms: the movement of the stars, the sun, the winds, and wave motion (swell and reflected waves), as well as underwater lightning phenomena, and the movements of birds and other marine animals. Learning was based on mnemonics involving the repetition of rhymes, songs, and dances to memorize the list of islands and the trajectories of the constellations; a further system involved adopting mental diagrams in the shape of a triggerfish or the constellation Crux, as a schematic metaphor for organizing the position of islands, star paths, and other key navigation data: "These skills must have included some competence in orienting and holding a course by reference to the stars, ocean swells, and winds, in dead reckoning, in sensing islands before they could be seen directly, and in incorporating newly found islands into some kind of cognitive chart" (ibid., p. 487). Magellano and later Forster immediately recognized the great skill of the master navigators among the Pacific islanders and attempted to transfer to a map the information dictated by a Tahitian, Tupaia, who accompanied the Endeavour expedition on the return leg of its voyage in 1769. Tupaia's map,[25] the best known example of indigenous geographical knowledge, represents the position of 74 islands in relation to Tahiti, within an area of 40 degrees longitude by 20 degrees latitude, but still remains difficult to decipher. According to the ethnologist Horatio Hale, the erroneous position of many islands on Forster's map, based on Tupaia's indications, is due to the fact that the Tahitan terms for south and north may have mistakenly been inverted in the transcription process (Mack, 2011).

On the Marshall Islands, but uniquely at this location, the skills of seafaring were learnt using marine "stick charts" made out of palm fibre, bamboo sticks, and shells that illustrated the refraction and intersection of wave motion. These "charts" were consulted prior to undertaking a sea voyage, but were also used by novice seafarers as an aid to memorization; they were not taken on board the sea craft: "The islanders' navigational 'map,' rather, was an intricate knowledge of the meaning to be gleaned from subtle signs in the seas and skies" (Suarez, 2004, p. 25). Capitan Otto Winkler, of the Imperial German Navy, in his study of navigation on the Marshall Islands, drew a diagram of the different categories of wave motion or swell observed around the islands. The diagram draws on local terminology to explain the pattern of wave refraction: *rilib* (swell from the east) and *kaelib* (swell from the west). In this detailed identification system for wave patterns, used, among other categories, by sailors to orient themselves on the open sea, we find terms to suggest that going wrong was a – likely not uncommon – possibility: “The northern and southern arms of a refracted *rilib* (eastern swell) are called, respectively, *rolok* (‘something lost’) and *nit in kot* (‘trap’ or ‘bird cage’), giving the sense that the navigator has overshot his target and must turn back” (Finney, 1998, p. 478).

There are numerous accounts of seafaring expeditions among the Pacific islands that went off course or ended tragically, to the extent that in Suarez’s view, a successful voyage was actually the exception to the rule: “The instance in which a lost vessel reached a new inhabitable island with its passengers alive must have been rare indeed, one lucky canoe for the many lost to the anonymity of the Pacific floor” (Suarez, 2004, p. 18). James Cook during a stopover at Tonga in 1777, observed that local navigators had difficulty in orienting themselves based on wave motion and wind when the sun and stars were hidden from sight: “they are then bewildered, frequently miss their intended port, and are never heard of more”. However, the islanders’ stories of their people being driven to other islands “leads us to infer, that those not heard of, are not always lost” (ibid., p. 19). Suarez reports various stories of seafarers getting lost among the Pacific islands and involuntary migrations between Melanesia and Polynesia:

> The plight of one lost Tuamotu fleet is documented in 1821. Three vessels with 150 people left the atoll of Anaa to greet the new king in Tahiti. The Anaa-Tahiti shuttle appears to have been a relatively common voyage, the navigator lining up his canoe on the sand straight for Matavia Point on Tahiti. Launching the craft at sunset with eyes fixed on bright star on the western horizon, a succession of stars would appear throughout the night, a living chart to illuminate the course to those privy to its hard-won secrets. Two days into the voyage, a gale separated the fleet and drove it back. One vessel, a 30-foot double canoe, reclaimed its course but was then greeted by another storm which sent it far into windless seas, where it was becalmed for more than a week under the hot, rainless sky. The two weeks’ provisions were soon exhausted, and of the 48 people on this vessel – 23 men, 15 women, and 10 children – 17 died.
>
> (ibid., p. 18)

After many other misadventures, only a few boats from the original party made it to the island of Vanavana, 800 km distant from the island they had set out from, and double the distance they would have needed to travel to reach Tahiti.

Whether the settlement of the Pacific, the ocean defined by Magellano as “so vast that the human mind can scarcely grasp it” (Finney, 1998, p. 445), was “deliberate or accidental” (Suarez, 2004), remains an open question that may never be answered. In his monograph *Ancient Voyagers in the Pacific*

(1957) and its revised edition *Ancient Voyagers in Polynesia* (1963), Andrew Sharp[26] put forward the theory of accidental drifting and critiqued the theory of the great Oceanic school of navigation, which he suggested was a romantic myth. In Sharp's opinion, the Pacific islands were mainly settled due to chance and loss of orientation on the part of seafarers rather than to planned voyages of discovery. Polynesia

> Was first discovered and occupied by sporadic accidental voyagers who were driven off course into the eastern Pacific by westerly storms. Settling on the islands they had discovered by chance, their numbers slowly increased. Then occasional voyagers from these islands became lost at sea and some of them also were lucky enough to make accidental landfalls and find new homes. The primary accidental settlement of Polynesia was from west to east, but the subsequent accidental voyagers might have been carried in almost any direction depending upon winds, currents, and seasonal conditions.
>
> (Luomala, 1957, pp. 776–777)

Sharp's hypothesis,[27] which has been invoked to explain the populating of many different areas of the world, on the one hand "calls attention to a factor too often overlooked" (Luomala, 1958, p. 778), but on the other, has been the object of hot controversy and accusations of Eurocentrism: "Andrew Sharp's contention that Pacific Islanders could have only reached remote islands accidentally is an excellent example of how ethnocentricities derived from these attitudes have structured theoretical modelling of seafaring and migration in Oceania (however short-lived)" (Fitzpatrick, 2007, p. 79). The seafaring tradition of Oceania is still alive today and its origins have been explored both at the theoretical level in fields ranging from archaeology to linguistics and ethnology and via experimental voyages in traditional craft (see Gladwin, 1970; Lewis, 1972; Finney, 1979, 1994, 2003; Kyselka, 1987, etc.).

Adrift

> There is a pleasure in the pathless woods,
> There is a rapture on the lonely shore,
> There is society, where none intrudes,
> By the deep Sea, and music in its roar
>
> (G. G. Byron, *Childe Harold's Pilgrimage*,Canto IV, Verse 178, 1812–16)

The myth of the lost island and its mysterious appeal has generated utopian and dystopian imaginaries in all eras: "The metaphoric deployment of 'island' is, in fact, so enduring, all-pervading and commonplace that a case could reasonably be made for it as *the* central metaphor within western discourse" (Hay, 2006, p. 26, emphasis in original). Disorientation is integral to the myth of the mysterious, unknown, and often fog-shrouded island, and contributes to its fascinating ambiguity: "The island seems to have a tenacious hold on the human imagination. […] Its importance lies in the imaginative realm. Many of the world's cosmogonies […] begin with the watery chaos: land, when it appears, is necessarily an island" (Tuan, 1974, p. 118). The theme of the island is popular in European culture, and is often developed in art and literature, from Thomas Moore to Shakespeare to Jules Verne: "Utopian fictions, we might say, are a form of poetic geography that function as a counterdiscourse to the positivistic geography based on measurement and the diagrammatic assembly of data" (Brayton, 2012). In Deleuze's terms, the island is not a geographical place, but an enchanted symbol of movement:

> The élan of humans that lead them toward islands takes up the double movement that produces islands in themselves. To dream of islands (with all the attendant anguish and joy) is to dream of separating oneself, that one is already separated, far from continents, that one recreates, that one recommences. There were derived islands, but the island is also what one derived towards, and there were originary islands, but the island is also the origin, the radical and absolute origin.
>
> (Deleuze, 2004, p. X)

Phantom islands appeared on and disappeared from ancient maps, generated by the imagination or navigation and measurement errors. The best known of these include Atlantis, the lost continent, mentioned by Plato in *Timaeus* and *Critias*, and "Ultima Thule", the symbol of the far north in medieval geography: "The North is known as 'the most faraway land of Thule' (Ultima Thule, Θούλη, *Thoulē*) […] the North is a place located beyond the borders of the known world" (Minaeva and Holmquist, 2015, p. X). The island of St. Brendan, one of the points supposed to have been reached by the saint on his famous sixth-century voyage, appears in various positions on many ancient maps and atlases: as legend would have it, perhaps transported from one to the other by a giant whale. It featured on Ebstorf's planisphere up to 1234 with the label: "Lost island. St Brendan discovered it but nobody has found it since" (Corbella, 1996, p. 132). A more recent cartographic error is associated with the mutineers on the Bounty. The island of Pitcairn, found by Philip Carteret in 1767, was added to the map some 150 miles off its true position and this allowed the fugitives from the Bounty to settle there and to remain undisturbed for 19 years until the arrival of an American ship. The island was recognized as a British colony in 1838.

In the contemporary era, the mystery island has lost none of its appeal, remaining a powerful metaphor for both paradise and hell: "Islands have occupied such a powerful place in modern Western imagination that they lend themselves to sophisticated fantasy and mythology. Paradises, but also Gulags, are generally islands" (Baldacchino, 2005, pp. 247–248). The island continues to feature in literature, film, videogames, manga, comics, and role-playing games. Islands can thus become "outposts of aberrant exoticism, peopled by innocent and exuberant natives" or "background for the enactment of a male and heroic paean to colonialism, the subject of Robinsonnades that extend up to the present in the likes of Tom Hanks' movie *Castaway* or the TV blockbuster series *Lost*" (Baldacchino, 2012, p. 56). The wilderness, whether forest or sea, as a primeval medium in which to get lost, not only offers a spatial dimension but is the place in which human beings must learn to come to terms with their origins and with the different character of the wild, and accept radical changes with respect to their usual way of life. The imaginary of the forest and the desert island are joined by the tragedy of the shipwreck, another theme that has occupied the poetic imagination for centuries. The literary genre of *shipwreck literature* is transversal to many cultures, and historical periods, but is especially prominent in English language literature, fruit of a male-dominated and heroic colonial vision, from Defoe, to Poe, Tennyson and Stevenson, and to the *Life of Pi* by Yann Martel. More recently *Adrift: Seventy-Six Days Lost at Sea* (revisited by the TV series *I Shouldn't Be Alive*) tells of the shipwreck and survival on a raft in the middle of the Atlantic Ocean of lone sailor Steve Callaham in 1982. Yet we read in Mentz that:

> Shipwreck makes providential history uncomfortable and disorienting. In place of timeworn understanding of modernity as displacement or disenchantment, shipwreck advances the more disorderly image of catastrophic accumulation. […] This fragmented world, like the edge of the sea, combines "order and chaos" and its development always remains to some extent "illegible". Early modern shipwreck narratives suggest that all waters […] share this potential to scatter sailors and their histories.
>
> (Mentz, 2013, p. 82)

Notes

1 Part of this chapter, here revised and extended, was previously published in Schmidt di Friedberg (2010).

2 A classic literary example are the caves in Mark Twain's *Tom Sawyer.*

3 On getting lost in a snow storm, see *Master and Man* by Leo Tolstoy (1895), and *Aus dem Bairischen Walde* by Adalbert Stifter (1867).

4 Humphrey reports, in relation to her study of the landscape of disorientation in Inner Mongolia: "Perhaps partly because the mountains are so dangerous and so bemusing with their twisting defiles leading to blank cliffs, 'losing the way' is an everyday preoccupation of Mongols. However, causing people to lose the way was also attributed by our respondents to graves and ghosts" (Humphrey, 2001, p. 63).

5 On contemporary understandings of wilderness, see Cronon (1995).

6 Descartes' argument here does not necessarily represent sound practical advice. Primo Levi, on his way home from Auschwitz concentration camp, got lost in the forest near Starye Dorogi. He describes finding out to his cost that getting lost in the woods, a frightening and disconcerting experience, is not just a literary device used in fairy tales for children, but can happen in real life: "I had lost the north". Levi recalls how exhausted he became as he continued to wander through the forest for hours, increasingly fearful as night began to fall: "Then I decided to

set off straight ahead […] without stopping until I met the main road, […] a prey now to utter panic, to the age-old fear of the dark. The forest and the unknown". Finally, the whistle of a train guided him back onto the right path and he realized that he "had been going in the wrong direction" (Levi, 1995, pp. 144–145).

7 *The Divine Comedy of Dante Alighieri*, The Italian Text with a Translation in English Blank Verse and a Commentary by Courtney Langdon, Cambridge, Harvard University Press, 1918.

8 Dante speaks of Brunetto Latini with great respect, but represents him in Hell, in the Ring of the Sodomites.

9 Holloway (1981).

10 *fiadh*, a deer, Irish *fiadh*, Early Irish *fíad*, Old Irish *fiadach*, venatio, Welsh *gwydd*, Breton *guez, goez*, savage, *veido-s*, wild; Old High German *weide*, a hunt, German *weide*, pasturage, Norse *veiðr*, hunting; further is Gaelic fiodh, wood, English *wood*. Hence *fiadhaich*, wild (MacBain's Dictionary, An Etymological Dictionary of the Gaelic Language, Section 18, MacBain, Alexander Gairm Publications, 1982).

11 See also dell'Agnese (2015).

12 "Several studies report that children and teens are frequently afraid of the woods, although the source of such fears may be varied […] In one study, children were asked to name the 'scary' places in and around the suburban area where they lived, and forested areas were second only to haunted houses in scariness (interestingly, the same children also indicated interest in visiting these woods, although they rarely went there); […] Indeed, much anecdotal evidence from many quarters supports the conclusion that most children and many adults have apprehensions about entering the forest, especially alone. It's important to note, however, that the woods themselves are rarely the stimulus for such fears, but are feared for what objects or experiences may be found there. That is, people don't fear clumps of trees so much as the bears, wild dogs, and other dangerous animals, as well as the darkness, the ghosts, and the strangers lurking there, no matter how unrealistic these fears may be" (Hill, 1998, p. 14).

13 "It's growing late, it's getting cold. Never more will you leave these woods."

14 "Verordnung zur Schaffung von Nationalparken im Gebiet des Deutschen Reiches" (Decree for the creation of national parks within the German Reich): "Nothing binds our people to the land more than the beauty of the German countryside and the varied and particular nature of its animal and vegetable world" (Draft Decree, 4 September, 1939).

15 W. Schönichen, *Naturschutz in dritten Reich*, Berlin, 1934, and *Naturschutz als völkische und internationale Kulturaufgabe*, Jena, 1942.

16 If the world's net forest area continues to decline by 5.2 million hectares per year (the average net annual loss between 2000 and 2010), it will take 775 years to lose all of the world's forests (FAO, 2012, p. 16).

17 Japan currently has the highest rate of urbanization in the world, with a megalopolis that is home to over 60 per cent of the country's inhabitants: two-thirds of the population live on only 3 per cent of the land area (Knight, 1997).

18 The Japanese word *yama*, mountain, also means forest.

19 From the 1960s onwards, the country has become increasingly dependent on imports of wood from other parts of Asia, Siberia, and North and South America, thus making a major contribution to the deforestation of south-east Asia (Philippines, Malaysia, Indonesia).

20 The popularity of the place has been greatly stimulated by novels (beginning with Seichō Matsumoto's 1960 work, *Tower of Waves, Nami no tō*), newspaper and media coverage, and film, including: the recent *Jyukai – The Sea of Trees Behind Mount Fuji*, 2004; *Grave Halloween*, 2013; *The Sea of Trees*, 2015; *The Forest*, 2016.

21 If a man does not know to what port he is steering, no wind is favourable to him.

22 Neither the relationship between Toscanelli and Colombus nor the authenticity of Toscanelli's letter about Columbus' undertaking have been proven, and there are conflicting views on the matter (see COLOMBO, Cristoforo, Dizionario Biografico degli Italiani – Volume 27, 1982, by Marianne Mahn-Lot).

23 *Dead reckoning* or *path integration* is "the process by which a subject continuously monitors the route taken from a particular reference position so that he or she always knows his or her distance and direction from the reference point" (Etienne and others, in Golledge, 1999, p. 199).

24 On the Cook Islands, according to the missionary William Wyatt Gill, to ensure that a voyage would be successful and to obtain the assistance of Raka, god of the winds, magic rites were carried out, based on the use of a wind rose showing the directions and names of 32 winds marked with holes in the sides of a giant pumpkin (Wyatt, 1876).

25 It is not even clear who drew the original map (Finney, 1998. p. 447).

26 However, "The documentation by Lewis, Gladwin, and others of the superb skills of surviving navigators in the central Caroline Islands of Micronesia and a few other remote places, plus the realistic insights on navigation and sailing strategy provided by sailing Hokule'a over the long sea routes of Polynesia, served to demonstrate how stone-age sailors could have intentionally explored and colonized the Pacific and then traveled back and forth between distant outposts" (Finney, 1997, p. 131).

27 Sharp's theory has been confuted by the inhabitants of the Caroline Islands who, since 1970, have successfully travelled ancient routes among islands using traditional craft (see Polynesian Voyaging Society).

10 Lost in cyberspace and art[1]

Cyberspace

The expression "cyberspace" appeared for the first time in 1982, in a short story by American writer William Gibson entitled *Burning Chrome*. It became popular after Gibson used it again in his 1984 novel *Neuromancer*. Cyberspace, according to Gibson, is: "a consensual hallucination experienced daily by billions of legitimate operators, in every nation [...] A graphic representation of data abstracted from the banks of every computer in the human system. Unthinkable complexity. Lines of light ranged in the non-space of the mind, clusters and constellations of data. Like city lights, receding" (Gibson, 1984, pp. 28–29). In the 1990s, the term coined by Gibson was extended to cover all that concerns the use of Internet and Internet culture. Thus, cyberspace has come to represent a territory that encompasses multiple sensory dimensions and is highly dynamic, being continuously reshaped, constructed, and explored by its tens of millions of users: "The network as such will remain an unknowable system – an invisible territory, maps of which we continually redraw by surfing the net" (Harris, 1996, p. 1).

The availability of increasingly more sophisticated and user-friendly computers and information systems currently allows millions of people to interconnect in the virtual world and spend hours online daily. However, "travelling" on the Net can clash with our usual way of experiencing movement and orienting our body and senses in the real world. The world of Internet and virtual reality can therefore make us feel disoriented: "The process by which media worlds create orientational images and deterritorial identities thus also produces disorientation. These always become evident wherever resistance emerges to adaptation to an imaginary life world in neutral space" (Hornbacher, 2010, p. 97). Another feature of cyberspace that can disorient us is the vast amount of information accumulated inside it, whose sheer extent and apparently chaotic organization can give rise to a sense of impotence. In the words of "information architect" Richard Saul Wurman:

> There is a tsunami of data that is crashing onto the beaches of the civilised world. This is a tidal wave of unrelated, growing data formed in bits

> and bytes, coming in an unorganised, uncontrolled, incoherent cacophony of foam. [...] The tsunami is a wall of data on tape, on disks, on paper, sent by streams of light. Faster and faster, more and more and more.
>
> (Wurman, 1997, cit. in Engeli, 2000, p. 30)

The spatial metaphor of the Web

Movement through cyberspace and the associated sense of disorientation are primarily described using spatial metaphors; cyberspace is a network, a global village, we may navigate, surf, or travel on its information highways until we reach a safe port or home, a starting point or destination; these metaphors are never random and their use contributes to the construction of virtual reality:

> Thus, metaphor may allow us to avoid or at least to mitigate the disorientation issues that arise when we consult hypermedia. These appear when the user has the feeling of being lost inside the system; he is no longer able to identify his position within the overall structure, retrace the path that has led him to this point, discern the possibilities now open to him, or choose a destination and work out a path towards a point that he knows to exist.
>
> (Collard, 2006, p. 86)

In cyberspace, the metaphor is a linguistic process that can make us feel less lost as we grapple with hypermedia; it is an attempt to introduce order. Conceptualizing the virtual world as a space allows us to change position within it without physically moving; it helps us to attribute structure to chaos and makes hyperspace familiar and imaginable. For Paul Ricoeur, the metaphor is: "The rhetorical process by which discourse unleashes the power that certain fictions have to redescribe reality" (Ricoeur, 2004, p. 5). The virtual world may be explained and encountered through metaphor, by translating it into the geographical language of place and space, a space that is topological rather than Euclidean:

> All the metaphors used by Gibson, as well as the labyrinth and city metaphor, present the Internet as a complex, well-articulated space, which needs to be interpreted (labyrinth), and mapped (city), because it is formed by a number of different places (such as a ferns, a forest, and clusters), which could even possess an esoteric meaning (pyramid, constellation, and labyrinth).
>
> (Bittarello, 2009, p. 4)

Thus, for cyberspace also, the labyrinth remains a dominant metaphor, though taking on new spatial forms.

The virtual labyrinth of cyberspace

The new form of labyrinth is the rhizome, a complex semantic model that replaces the concept of the tree-shaped labyrinth based on a hierarchical and linear organization of flows and information: "unlike trees or their roots, the rhizome connects any point to any other point, and its traits are not necessarily linked to traits of the same nature; it brings into play very different regimes of signs, and even nonsign states. […]. In contrast to centered (even polycentric) systems with hierarchical modes of communication and preestablished paths, the rhizome is an acentered, nonhierarchical, nonsignifying system" (Deleuze and Guattari, 1987, p. 21). The rhizome, in other words, is a "third kind" of labyrinth, to be added to the classical unicursal labyrinth whose single coiled path leads from the entrance to the centre, and the Mannerist labyrinth, maze, or *Irrweg* that offers many alternative paths and dead ends (see Chapter 7, § 7.6). As observed by Umberto Eco:

> Every point of the rhizome can be connected to any other point; it is said that in the rhizome there are no points or positions, only lines; this characteristic, however, is doubtful, because every intersection of two lines makes it possible to identify a point; the rhizome can be broken and reconnected at any point; the rhizome is antigenealogical (it is not an hierarchized tree); if the rhizome had an outside, with that outside it could produce another rhizome, therefore it has neither an inside nor an outside; the rhizome can be taken to pieces and inverted; it is susceptible to modification; […] a global description of the rhizome is not possible, either in time or in space; the rhizome justifies and encourages contradictions; if every one of its nodes can be connected with every other node, from every node we can reach all the other nodes, but loops can also occur; only local descriptions of the rhizome are possible.
>
> (Eco, 2014, pp. 54–55)

Hence the third kind of labyrinth is a network, all points in which may be connected to all other points. In the virtual era, the labyrinth yet again becomes a powerful metaphor, this time for interpreting cyberspace: the information technology labyrinth is in a continuous state of transformation and has neither a centre, starting or end point: "Cyberspace labyrinth belongs to another class, another typology. It maintains characteristics of the first kind, but it goes beyond. […] A rhizome can be connected in different directions and from each of its points – so [too] WWW" (Leão, 2000, p. 5). The rhizomatic labyrinth metaphor allows us to imagine the World Wide Web as a multidimensional network.

Yet is metaphor really an appropriate tool for conceptualizing cyberspace? According to Anders: "The trouble with the spatial metaphor in computers is that it can limit our understanding of cyberspace" (Anders, 1996, p. 4). Thus, it is crucial to ask whether cyberspace may truly be considered a space at all, given that our movement within it is purely metaphorical: "How can an assemblage of cables, routers, and servers be a space?" (Cohen, 2007, p. 227) enquires Cohen, in the context of a study on the role of metaphor in legal discourse and cyberlaws. Key to the notion of cyberspace, she goes on to argue, are users and their capacity to produce cyberspace itself: "'Cyberspace' is most usefully understood as connected to and subsumed within an emerging networked space that is inhabited by real, embodied users and that is apprehended through experience" (ibid., p. 255). When I "navigate" from my chair, the virtual splits off from the real causing me to lose contact with reality; Paul Virilio has argued that the "here" and "now" are called into question in the virtual dimension. In cyberspace, I can carry out financial transactions, purchase goods, search for information, and entertain social and romantic relationships; yet cyberspace is not a parallel world duplicating the real one, but a qualitatively different entity, without borders.

The world of Internet is subject to a process of constant transformation, within a temporal dynamics dominated by speed, and the phenomena of immediacy and instantaneity, meaning that everything ages more rapidly. In Mitchell's view, not only does Internet not observe the formal rules of real space but it is strongly anti-spatial: "The Net negates geometry. While it does have a definite topology of computational nodes and radiating boulevards for bits, and while the locations of the nodes and links can be plotted on plans to produce surprisingly Haussmann-like diagrams, it is fundamentally and profoundly antispatial. [...] You can find things in it without knowing where they are. The Net is ambient – nowhere in particular but everywhere at once" (Mitchell, 1996, p. 8). The space of hypertext, in turn, is never linear, as the contents are blended with instructions for the user and links to further information, which is presented in a way that is fragmentary and often perceived as chaotic by the non-expert. Virilio has conceptualized cyberspace as a new form of perspective:

> It does not coincide with the audio-visual perspective which we already know. It is a fully new perspective, free of any previous reference [...] with the build-up of information superhighways we are facing a new phenomenon: loss of orientation. A fundamental loss of orientation complementing and concluding the societal liberalization and the deregulation of financial markets whose nefarious effects are well-known. A duplication of sensible reality, into reality and virtuality, is in the making. A stereo-reality of sorts threatens. A total loss of the bearings of the individual looms large. To exist, is to exist in situ, here and now, hic et nunc. This is precisely what is being threatened by cyberspace and instantaneous, globalized information flows.
>
> (Virilio, 1995, p. 28)

When I "surf" the Internet, my body remains physically bound to my desk, while "I" – or at least, an aspect of me – "travel" in another dimension with a logic and geography of its own that are no less real because immaterial (Schroer, 2006, p. 263). In cyberspace, boundaries become irrelevant: and not only geographical boundaries but also those based on age, gender, ethnicity, social status, and identity, all information that may be withheld, modified or only partially revealed: "Apparently we are not just looking at a despatialization of society. Cyberspace appears to bring the society without boundaries far closer to us" (ibid., p. 252). Nonetheless, from this hyper-reality, which appears to be unlimited, there gradually emerges a system of parcellizations, consisting of barriers and walls, controlled by filters, firewalls, passwords, *no-go-areas* and pay sites that interrupt navigation: "You don't get to go just anywhere in a city, and the same is true of cyberspace. In both domains, barriers and thresholds play crucial roles. [...] Spatial divisions and access-control devices are carefully deployed to organize places into hierarchies grading from completely public to utterly private" (Mitchell, 1996, p. 122).

The world of the web, similarly to the real world, appears to be structured according to logics of power that define property and access rights; the outcome is a network of *gated communities* that reproduce social processes of inclusion and exclusion found in the real world, further heightening our sense of disorientation:

> In sum, we are dealing with the dynamic ongoing construction (*Grenzenaufbau*) and dismantling (*Grenzenabbau*) of borders, violation (*Entgrenzung*) and limitation (*Begrenzung*), despatialization (*Enträumlichung*) and spatialization (*Verräumlichung*). Crucial to and typical of our "spatial era" [...] is precisely the fact that solid and liquid, land and sea, real and virtual, Internet and Intranet can no longer be conceived as mutually exclusive conditions but exist contemporarily, one beside the other.
>
> (Schroer, 2006, p. 274, italics added)

Cyborgs

The virtual world gives an illusion of space that is based on metaphor and therefore purely symbolic. Hence its ambiguous relationship with the real world, which it is both detached from and bound up with: "Cyberspace extends our cognitive space. [...] Cyberspace anomalies *express* the problematic relationship between cyberspaces and the physical world. They occur when the spatial illusion breaks down and cyberspace betrays its nature. Cyberspace – a purely symbolic space – often has no foundation in the physical" (Anders, 1998, pp. 88–89). In the "augmented space" (Manovich, 2006) of virtual reality, physical space acquires new coordinates, dematerializes and takes on novel dimensions that transcend the laws of physics; in the field of architecture, cybrids are produced. A *cybrid* is: "An environment or artifact that incorporates both physical and cyberspaces. The relationship between these two types of space may be distinct, overlapping or congruent with one another" (Anders, 1999, p. 195).

As we travel in cyberspace, we may immerse ourselves in a virtual environment in which our sensory faculties are stimulated via peripheral devices, such as HMDs or *head-mounted displays*, visors or glasses for stereoscopic vision, datagloves, data suits, or simply a mouse or joystick. Experiencing the virtual world and exploring its spaces via visors and joysticks has been imagined and depicted by writers of fiction. One such account is that of Gibson in *Mona Lisa Overdrive*: "Her head filled with a torrent of white noise. Her fingers found a random second stud and she was catapulted through the static-wall, into cluttered vastness, the notional void of cyberspace, the bright grid of the matrix ranged around her like an infinite cage" (Gibson, 1988, p. 56).

Reflecting on virtual space means reflecting on the body and its augmented cognitive capacities: "The body takes on a prosthetic virtual life of its own, a life then seen by a disembodied spectator's viewpoint; and from this situated spot, the projected body image reveals the visual inside of the embodiedness of living biological infrastructure. From the outside, the VR-user lives an uncanny relation to the body-image; it becomes a double, but with the further twist that this doubling relation itself finds objectified form. The VR-user doesn't so much experience the body as other as the very process of othering" (Harris, 1996, p. 1). In order to describe this fusion of bodies and machines, in 1960 two NASA scientists, Manfred Clynes and Nathan Kline, coined the term "cyborg": "The cyborg deliberately incorporates exogenous components extending the self-regulatory control function of the organism in order to adapt it to new environments" (Clynes and Kline, 1960, p. 27).

The provocatory image of a posthuman world that is biotechnological, visual, virtual, and inhabited by body-machine hybrids is at the heart of Donna Haraway's Cyborg Manifesto:

> A cyborg is a cybernetic organism, a hybrid of machine and organism, a creature of social reality as well as a creature of fiction. Social reality is lived social relations, our most important political construction, a world-changing fiction. [...] This experience is a fiction and fact of the most crucial, political kind. Liberation rests on the construction of the consciousness, the imaginative apprehension, of oppression, and so of possibility.
>
> (Haraway, 1991, p. 149)

In the hybrid world of the cyborg, the metaphor of the labyrinth comes to the fore yet again: "Cyborg imagery can suggest a way out of the maze of dualisms in which we have explained our bodies and our tools to ourselves. [...]. It means both building and destroying machines, identities, categories, relationships, space stories. Though both are bound in the spiral dance, I would rather be a cyborg than a goddess" (ibid., p. 181).

Through interfaces with virtual reality, the body expands towards novel horizons at which new technologies challenge traditional notions of the human, non-human, artificial and natural and their increasingly fuzzy boundaries (Agamben, 2001; Whatmore, 2002; Thrift, 2008, etc.). Non-representational theory (NRT), or what Lorimer (2005) has referred to as "more than representational theory", explores the essence of the body and its presumed naturalness in the virtual era: "Non Representational Theory seeks to better cope with our self-evidently more-than-human, more-than-textual multisensual world (Lorimer, 2005, p. 83). Technology continuously gives rise to instruments for enhancing our bodily capacities and minimizing its imperfections: "The human sensorium is constantly being re-invented as the body continually adds parts in to itself" (Thrift, 2008, p. 2). Artificial prostheses, such as GPS, mobile phones, and GIS, guide us through a technological world and virtual space: "A series of prostheses which routinely offer cognitive assistance and which do much of the work of navigation automatically" (ibid.).

We rely on satellite technologies to get around, while having a traditional sense of orientation appears to be of no practical use in everyday life. On the other hand, the virtual world, now fully part of our lives, confronts us with an overload of information:

> The term "disorientation" has become part of our everyday language. While information technologies are constantly being perfected, the citizen of our mediatic postdemocracies appears to feel disoriented for a host of reasons. [...] It is undeniable today that the human being delegates responsibility for spatial orientation to a machine, just as in the past responsibility was handed over first to the clock and then to the watch. One day, other new instruments will solve all of our navigation issues in their entirety. What will be the consequence for the environment of this systematic assistance and, more generally, on our ability to avoid getting lost?
>
> (Baur, 2009, intro.)

Does it still make sense to invest in developing sensory skills for interpreting real space and learning how to orient ourselves in it? In Thrift's opinion: "The sense of direction will become a given. It will no longer be something that has to be considered" (Thrift, 2008). The new technologies offer us instruments that are increasingly accurate, meaning that we can no longer relate to a space by getting lost in it or wandering about it with no fixed destination in mind (*flâneur*). Yet we suffer from a heightened sense of disorientation, lost in a sea of data: "all travellers on the Web have experienced the disorientation of finding themselves in a strange location, in-between the familiar and the unknown, and wondering 'Where am I?' 'Why am I here?' 'What does this mean?' The key element underlying this network of choices is the *link*. [...] The sense of being 'lost', then, is both a problem of having arrived in an unfamiliar location, and an *aporia* of unknown signification" (Burbules, 1997, p. 1). New technologies have radically modified our lifestyles, our leisure time activities, and our social relations. Kajmaku concludes that: "The labyrinth of the third millennium are the social media" (Kajmaku, 2015–2016, p. 33).

Mapping cyberspace

If we wish to go on viewing cyberspace as a space, then it should be possible to translate it into a cartographic language (Bittarello, 2009). However, while some elements of information networks, such as their infrastructure, users, set of connections or providers, may easily be represented on a conventional topographical map, others are highly complex: "This is because the spatial geometries of cyberspace are very complex, often fast-changing, and socially produced. Cyberspace offers worlds that, at first, often seem contiguous with geographic space, yet on further inspection it becomes clear that the space–time laws of physics have little meaning online" (Dodge and Kitchin, 2001, p. 3). Dodge and Kitchin, in their *Atlas of Cyberspace*, present a collection of over one hundred maps or mapping projects aimed at representing cyberspace as a whole, including its constituent infrastructures. For, in practice, cyberspace is the product of a vast range of IT instruments that are programmed by technicians and millions of users. It is thus a "purely relational" space: "Cyberspace consists of many different media, all of which are constructions; that is, they are not natural but solely the production of their designers and, in many cases, users. They only adopt the formal qualities of geographic (Euclidean) space if explicitly programed to do so; and, indeed, many media – such as email – have severely limited spatial qualities" (Dodge and Kitchin, 2001, p. 3).

The attempt to map cyberspace can take many different forms: a spatial visualization may be obtained by transferring onto a map the system of infrastructures making it up, its user network, data, and traffic flows. A leading example is the *Great Circle Map*, designed by artist Gill MacDonald for the British *Cable and Wireless Company* (C&W) in 1945: a map decorated with illustrations of the telecommunications system of the period, with the related technologies, and a set of curved lines indicating C&W's international connections. The commercial aim of this map was to represent and show off the company's global connections network and to highlight the geopolitical leadership of Great Britain, situated at the centre of a polar projection, given its role as "hub of the world" and its relations with the colonial areas (in red on the map).

Other cartographic projects attempt to spatially represent what is probably the most extraordinary aspect of cyberspace, that is to say, the network of information forming it, a highly dynamic and changeable system in continuous expansion:

> Trying to apply traditional mapping techniques to such spaces is all but impossible, because they often break two of the fundamental conventions

> that underlie Western cartography: first, that space is continuous and ordered; and second, that the map is not the territory but rather a representation of it. In many cases, such as maps of websites, the site becomes the map; territory and representation become one and the same.
>
> (ibid.)

Huge and constantly evolving data flows are produced and cancelled in a very short time, thanks to the participation of millions of users via email, chatrooms, MUD (*multi-user dungeons* or *domains*), virtual worlds, social networks, and games. Metaphorical language and spatialization make information easier for users to grasp than do infinite rows of numbers and data: "Cyberspace, although not spaceless as hypothesized by some analysts, does not have a static, easily measurable space–time geometry. This is because cyberspace is entirely socially produced – its form and structure are dematerialized and determined by its creators. Space–time, then, is discontinuous" (ibid., p. 75). The architecture of cyberspace can also issue from the imagination of artists in the fields of fiction, film, graphics, and design.

Artistic visualizations of the Web can act on the imagination of users, producing new and untidy representations of Internet that subvert the order carefully constructed by its creators via browsers such as Netscape Navigator or Internet Explore. An example of such a visualization is *The Web Stalker* (London, 1997)[2] by the artists Matthew Fuller, Simon Pope, and Colin Green, who set out to provide an alternative browser that overturns the conventional representation of webpages and redraws the Web, using simple elements to portray its complexity (ibid.). The work "1:1 – Every" by artist and researcher Lisa Jevbratt is a new form of interface, a map of IP (*Internet protocol*) addresses at the 1:1 scale that visualizes each individual address as a pixel within a large multi-coloured image (the colour of each pixel is related to the numbers making up the address it represents); the resulting image is a snapshot of one dimension of the overall Web, a set of coloured bands that looks like a TV screen disrupted by interference. Jevbratt commented that the interfaces in the 1:1 project: "are not maps of the Web but are, in some sense, the Web. They are super-realistic and yet function in ways images could not function in any other environment or time. They are a new kind of image of the Web, and they are a new kind of image" (Jevbratt, in Dodge and Kitchin, 2001, p. 249).

Within the "shifting boundaries" of cyberspace, where the real blends with the virtual, like avatars, we can live other lives in other worlds. An article in *National Geographic* reports on the experience of hundreds of people seeking safety on rooftops or in flight during a flood caused by climate change and affecting different locations – London, the Netherlands, Ibiza: we are in the virtual world of Second Life and, as avatars, we must react to the catastrophe.[3] Millions of users of Second Life and *World of Warcraft* (13 million active subscribers) move around and meet in virtual worlds endowed with a physical and cartographic dimension of their own, and inhabited by lifeforms of all kinds. Furthermore, numerous MMORPGs (*Massively multiplayer online role-playing games*) operate within the real economy, providing the opportunity to make or lose money and to establish "real" social relations. Hours of navigation are required to get to the imaginary planet of Azeroth, in *World of Warcraft*, and players may get lost along the way. On the other hand, it is possible to explore the virtual world of Azeroth using Google Maps API. The cartography of part of the continent of Azeroth is extraordinarily precise and detailed and all the relevant data is stored in the WoW database (2007).

The rules of perspective

Multicoloured images of cyberspace and the extraordinary capacity of virtual reality to transform our surroundings lead us to the last of the themes addressed in this book in which I have set out to provide an overview of the phenomenon of disorientation: the world of art. Long before the era of computers, poets, writers, painters, musicians, and sculptors had constructed fantasy worlds that challenged our habits and schemas of perception and looking. Art, in any possible expressive form – painting, plastic arts, photography, film, literature, music – anticipated cyberspace by giving access to new spatial structures, both real and imaginary, different to those familiar to us and consequently often strongly disorienting: "The artist brings back from the chaos *varieties* that no longer constitute a reproduction of the sensory in the organ but set up a being of the sensory, a being of sensation, on an anorganic plane of composition that is able to restore the infinite" (Deleuze and Guattari, 1994, pp. 202–203). Deleuze and Guattari go on to elaborate on this concept, stating that: "Art is not chaos but a composition of chaos that yields the vision or sensation, so that it constitutes, as Joyce says, a chaosmos, a composed chaos – neither foreseen nor preconceived. Art transforms chaotic variability into *chaoid* variety, [...]. Art struggles with chaos but it does so in order to render it sensory, even through the most charming character, the most enchanted landscape" (ibid., pp. 204–205).

In the West, from the Renaissance onwards, the rules of perspective have provided the basis for a shared model of the organization of vision: "In the history of art, perspective is generally considered as a factor of realism that renders the third dimension. In fact, it is above all an artifice that can serve all sorts of ends depending on one's vision of the work one is setting out to create" (Baltrušaitis, 1955, p. 13). Franco Farinelli has argued that in fifteenth-century Florence, the vanishing point of linear perspective[4] gave rise to the invention of space, and its mathematical rules: "Florentine perspective, the spatial agent of modernity, radically reconfigured not only the world of objects but specularly (a truly appropriate choice of word) also the nature of the subject, the opposite end of the knowing process" (Farinelli, 2009a, p. 40). Distinct from place, or "a part of the earth's surface that is equivalent to no other, that cannot be exchanged with others without thereby changing everything" (Farinelli, 2003, p. 11), space is understood: "in the Ptolemaic sense of the standard linear interval between two geometric points [...] in relation to which each use value, that is to say each place, is destined to disappear" (Farinelli, 2009a, p. 29). In Farinelli's view, space was invented in Florence: "Under that formidable monument, which is the most important not just in Florence, but in the whole of modernity and the whole of the Western world, the *Portico degli Innocenti*, built by Filippo Brunelleschi [...] for the sole purpose of showing his fellow citizens what space was" (Farinelli, 2009b, pp.3–4). Thus, the perspective built into the Portico degli Innocenti revolutionized the acts of looking and perception and replaced the notion of place with that of space:

> That space which is a completely inhuman and artificial way of conceptualizing reality, because under the Portico degli Innocenti both the classical and the medieval worlds came to an end: it was enough to stand where Filippo Brunelleschi wanted people to stand, that is to say, at one of the two false doorways at either end of the portico and look straight ahead [...] Perspective [...] presupposes an immobile observer [...] observers are obliged to look at everything at once, their gaze cannot be arrested because it is sucked in – so to speak – by the vanishing point, as noted by Leon Battista Alberti, an infinite vacuum opens up before them, empty and infinite space.
>
> (ibid.)

The objective representation of empirical space, based on the vanishing point of linear perspective, produces a stable and measurable horizon with respect to which we may orient ourselves. In the words of German artist and filmmaker Hito Steyerl: "Disorientation is a temporal trope [...] because it marks the dissolution of a specific history of the horizon. The historical emergence of a horizon with a single vanishing point in Renaissance art, mirrors and produces the subjectivity of the individual observer, and marks the emergence of orientation as an ideological project. It produces a space that defined by linear perspective is calculable, navigable, and predictable" (Steyerl, 2012, p. 18). Thus, orientation becomes an "ideological agenda" within a particular historical and political concept: "The linear space produced by the stabilisation of the horizon produces in turn linear time and linear progress: it constitutes, in other words, a vital addition to the 'tool kit for enabling Western dominance, and the dominance of its concepts' [...], furnishing the measuring stick for producing the difference through which the West marks its Others" (Martin and Rosello, 2016, p. 8).

The rules of modern perspective organize our world view but the artist can play around with them to confuse and disturb us: anamorphosis, a term coined in the 1600s to describe a technique that had long existed, inverts the elements and principles of perspective:

> It projects forms outside of themselves instead of reducing them to their visible limits and breaks them down so that they may be recomposed at a later time, when they are seen from a certain viewpoint. The procedure became known as a special technique, but it contains a poetics of abstraction, a powerful optical illusion mechanism and a philosophy of artificial reality. It is a rebus, monster and prodigy.
>
> (Baltrušaitis, 1955, p. 15)

Whether pursued for its novelty value, to showcase virtuosity, or as a vision of the world informed by the sciences of the occult, anamorphosis disorients. In the 1700s and 1800s, it was attributed with all kinds of mysterious meanings: "This reminds us that in all perspectives, whether optical perversion or disturbing *trompe-l'oeil*, there is always an element of surprise and of artifice. Indeed, perspective does not present itself as a science of reality, but as a tool for creating hallucinations" (ibid., p. 14). Maurits Cornelis Escher explores new imaginary spaces using optical effects and inverted perspectives that disorient on account of both their visual ambiguity and their ambiguity of meaning.[5] In his flights of stairs and impossible landscapes ("Day and Night", 1938; "House of Stairs", 1951; "Print Gallery", 1956, "Relativity", 1956), geometry and perspective are exploited to create illusions based on relations between the concave and the convex, while plays of light and shadow invert the usual relationship between the inside and outside of the figures, provoking a sense of vertigo. The spectator's eye is trapped in a labyrinth with no way out, forced to continuously range through an infinite space of possibilities, without any identifiable end or beginning.

Art as a form of transgression

In the early twentieth century, the world of art too found itself grappling with the devastating experience of the First World War and so the search got underway for new solutions and expressive techniques to free creativity, break with the constraints of perspective and orientation[6] and open up the mind, based on: "a rejection of the idea that art must cling to the representation of an everyday visible reality" (Smith, 1998, p. 400). The Surrealists violated the barriers of rational thought to explore the world of dreams and the unconscious. Transgression and the overturning of established conventions and patterns were now the key values inspiring art: "SURREALISM, *n.* Psychic automatism in its pure state, by which one proposes to express – verbally, by means of the written word, or in any other manner – the actual functioning of thought. Dictated by thought, in the absence of any control exercised by reason, exempt from any aesthetic or moral concern" (Breton, 1969, p. 26). De Chirico's "Pittura Metafisica" (metaphysical painting) of desolate landscapes, squares, and spectral pieces of architecture broke with the conventions of perspective, giving access to new temporal and spatial dimensions. Salvador Dalì drew on anamorphosis, fractal progressions and geometric illusions in his works in order to subvert the impressions received by the senses and create visual illusions from the world of dreams. He also ventured into the medium of film. His disturbing short *Un Chien Andalou* (1929), made with film director Luis Buñuel, engages the spectator in the labyrinthine, aggressive and discontinuous narrative of the two artists' oneiric imagination. Disorientation is a key theme in the film: the impossible images, and the victim-protagonist's wandering and fleeing only to find himself back in the same kind of place as before, reflect the structural ambiguity of a work that: "does not observe consistency of narrative time, place, character, or semantic universe" (Hedges, 1983, p. 46). Even the editing of the film to produce a rapid sequence of continuously shifting images and free associations that disorient the viewer, becomes an instrument of subversion. Thus, major changes were underway in both the culture and the tools of vision:

> With the twentieth century, the further dismantling of linear perspective in a variety of areas began to take hold. Cinema supplements photography with the articulation of different temporal perspectives. Montage becomes a perfect device for destabilizing the observer's perspective and breaking down linear time. [...] Time and space are reimagined through quantum physics and the theory of relativity, while perception is reorganized by warfare, advertisement, and the conveyor belt.
>
> (Steyerl, 2012, p. 22)

In our contemporary – post-modern, post-human, and post-colonial – era, the very logic of modernity, based on linear perspective and a stable horizon, is giving way. The use of ever new technological prostheses is transforming our relationship with space and time as well as our visual experiences:

> Our sense of spatial and temporal orientation has changed dramatically in recent years, prompted by new technologies of surveillance, tracking, and targeting. One of the symptoms of this transformation is the growing importance of aerial views: overviews, Google Map views, satellite views. We are growing increasingly accustomed to what used to be called a God's-eye view.
>
> (ibid., p.14)

In the quest for new interpretative models for contemporary reality, Hito Steyerl falls back on disorientation: "This disorientation is partly due to the loss of a stable horizon. And with the loss of horizon also comes the departure of a stable paradigm of orientation, which has situated concepts of subject and object, of time and space, throughout modernity. In falling, the lines of the horizon shatter, twirl around, and superimpose" (ibid.).

Contemporary artists are faced with a continuous stream of new communication tools, advanced technologies in every field and the need to express themselves and win visibility in a global world; they find themselves playing a leading role in the richness and contradictions of multicultural scenarios, which provide space for a diversity of voices, challenge patterns of Western assimilation, and foster the emergence of local identity as well as the drama of war and migration. *Disorientation – Contemporary Arab Artists from the Middle East* was the title of an exhibition at the House of World Cultures in Berlin, 2003, repeated in Abu Dhabi in 2009–2010 under the title *Disorientation II, The Rise and Fall of Arab Cities.* The exhibition provided a space in which Arab writers, film makers, musicians, and visual artists from the Middle East could express themselves, particularly in relation to conflicts in the region. In seeking new instruments and models of representation, artists come up against challenges associated with the worlds of power, politics, and the market:

> The art field is a space of wild contradiction and phenomenal exploitation. […] a place of power mongering, speculation, financial engineering, and massive and crooked manipulation. But it is also a site of commonality, movement, energy, and desire. […] It's messy, embedded, troubled, irresistible. […]. Art is not outside politics, but politics resides within its production, its distribution, and its reception.
>
> (Steyerl, 2012, pp. 99–100)

Performance and transgression attempt to break with conventional forms of expression in order to challenge the notions of normality, appropriateness of place, and being out of place:

> Transgression is important because it breaks from "normality" and causes a questioning of that which was previously considered "natural", "assumed", and "taken for granted". Transgressions appear to be "against nature"; they disrupt the patterns and processes of normality and offend the subtle myths of consensus. These deviations from the dominant ideological norms serve to confuse and disorientate.
>
> (Cresswell, 1996, p. 26)

Cresswell analyzed the first graffiti that appeared in New York in the early 1970s and their offensiveness to the notion of "order":

> Graffiti flagrantly disturbs notions of order. It represents a disregard for order and, it seems to those who see it, a love of disorder – of anarchy, of things out of place […] Graffiti is not just "out of place" because it is misplaced figuration; its "otherness", is also connected to its assumed source, the ethnic minorities of urban New York. […] Dirt and madness, too, are often used metaphorically to describe the third world and its imagined inhabitants.
>
> (ibid., pp. 42–3).

Artists on the other hand, even the graffiti artist – "the chaotic untamed voice of the irrational […] rebellious, irrational, dirty, and irreducibly 'other.' […]

the insane spoiler who resists reason and introduces chaos" (ibid., p. 45) – operate within the incumbent logic of a global neoliberalist market, which is perennially and rapidly changing, and detains the power to impose or take over new fashions and to assimilate and sell even the most radical forms of expression. New York's graffiti is turned into "legitimate art" when, from the carriages of the subway train and the walls of outlying suburbs, it enters the galleries of SoHo, a "proper" place in which to exhibit art: "Graffiti is simultaneously repressed and commodified" (ibid., p. 52).

The art of disorientation

Outside of galleries and the "appropriate" places in which to exhibit, artists go out into the open, with nature and wild spaces becoming the new material to be modelled. *Land Art* originated in the United States at the end of the 1960s as a form of protest against post-industrial capitalism, the art gallery market, and bourgeois culture:[7] "Land artists fight against the notion that we can 'attain a perfect God's eye view of the world' in confronting us rather with works which challenge the notion of orientation. These works disorient our classical sense of perspective since they function like anamorphoses, changing according to our position in space" (Rigaud, 2012, pp. 30–31). Disorientation is a key element in many works of land art: "You begin your travels by being immediately lost" (Smithson, in Shapiro, 1995, p. 40), observed Robert Smithson, the creator of *Spiral Jetty* (1970), an enormous spiral-shaped sculpture that is sunk in the waters of the Great Salt Lake in Utah. This kind of work totally alters the spectator's relationship with the work of art, which extends through isolated and immense natural spaces, becomes ephemeral, and is explored through the movement of those who experience, traverse, and get lost in it: "Land artists disorient the viewers in creating a process of loss and ignorance where the sense of knowing space and geography seems to disappear behind a new form of aesthetic experience. Orientation, in these works, is primarily about making it 'through' a space, instead of establishing the proper spectatorial distance from which to view a space" (Rigaud, 2012, p. 3).

In many cases, the realization of land art pieces, which are often of gigantic dimensions, situated in isolated places and difficult to reach, demands major intervention in the landscape, which serves both as background and medium for the works: "Land artists seek to challenge the supremacy of vision and perspective by pressing us into situations of physical disorientation. The artist, as Smithson suggested, is a 'site-seer' […] who teaches us a different way of looking at the land. These works challenge each viewer individually by making him go through a ritualistic experience of the land. Viewers are made to experience physical or cognitive disorientation, which transforms them into explorers of the land, navigating between orientation and disorientation. Our experience of the works is no longer based on knowing but on finding" (ibid., p. 9). Many of these creations are continuously modified by the elements and the passage of time, and may even disappear completely leaving a trace only in video footage and photographs. Land Art poses an additional challenge to the geographical map. In commenting on his installation "Mono Lake-non site" Smithson remarked that:

> Maps are very elusive things. This map of Mono Lake is a map that tells you how to get nowhere. [...] The lake itself is a salt lake. If you look at the map, you'll see it in the shape of a margin – it has no center. [...] the site is the unfocused fringe where your mind loses its boundaries and a sense of the oceanic pervades, as it were. I like the idea of quiet catastrophes taking place.
>
> (Smithson, 1996, p. 249)

The elusive nature of the map is a theme that recurs in other works by Smithson who loves experimenting with cartography. As observed by Rigaud in relation to Untitled (Antarktis) Circular Map (1967) "which distorts the notion of orientation and of an absolute structure, turning the map into a labyrinth":

> The works call for a reappraisal of our position in space in relation to the land, our use of it, our right to it, the languages that have crossed it, the control that we exercise over it by knowing its name. Yet that control is suspended when the map is no longer an evocation of the land but the land itself; these artworks negate the idea that a map could give an objective sense of the land and rather suggest that the land must be sensed individually for the map to have any meaning.
>
> (Rigaud, 2012, p. 6)

The disorientation of contemporary artists can also arise from the dismay and vertigo of encountering a blank sheet or canvas, or a material that has yet to be modelled, from the difficulty of composing/decomposing reality in all its complexity or the fear of being unable to realize a work as they have planned it: "The labyrinth is the blank sheet", in the words of the painter Marco Benedetti.[8] Once more, disorientation is a metaphor for the current period, in which artists are continuously trying out new forms of expression and do not have set standards by which to orient themselves. *Smarrimenti* [Disorientations] is an exhibition of 28 graphic works inspired by the first tercet of Dante's Divine Comedy,[9] and executed by the Italian painter and illustrator, Guido Scarabottolo: "The Divine Comedy has engaged great illustrators for centuries, up to and including our own day. It is unthinkable that my work could measure up to theirs: and this is another reason to feel disoriented".[10] The visitor can freely wander among the artworks, which combine drawing and writing. Scarabottolo sought inspiration from Dantean disorientation: "So I took the liberty of 'reflecting' on my doubts concerning the work that I do and the current era. Though I am well over halfway along the path of life,

I identify with the gloomy wood and the state of being lost. So I thought it was better to invite my five readers to join me in this disoriented state and try to reflect with them, than try to dazzle a greater number of people with the usual mirrors."[11] Scarabottolo also argues that for him disorientation means experiencing "a time of unbelievable change in the media of cultural transmission", and the revolutionization of his artistic career by the advent of new technologies and digital language. Disorientation also comes into the difficulty of imagining the future and how his role as an artist may further evolve: "Personally, I experience these times with the conviction that every step we take leads us further into gloomy woods and, no matter how hard we try to rationalize the situation, disorientation – at least for me – is inevitable." Scarabottolo's final observation on the theme: "The true darkness, made of dazzling lights, is out there."[12]

(a) (b)

Image: Guido Scarabottolo, smarrimenti (disorientations) stamping ink on paper

Notes

1 Part of the material in this chapter, presented here in a revised and extended form, was previously published in Schmidt di Friedberg (2012, pp. 121–131).

2 http://bak.spc.org/iod. Accessed 10 April 2017.

3 "London, Tokyo Submerged by Rising Seas". In "Second Life", John Roach, *National Geographic News*, 4 April, 2007.

4 For Farinelli, perspective bore huge innovative force that would influence the whole of modernity: "Perspective, stated simply, is a trick for drawing something on a flat surface while giving an illusion of depth. I see these parallel lines imperceptibly

converging in such a way that, if infinitely extended towards the vanishing point, they merge. And so, for the first time in Western human history, I need to decide whether to believe my eyes or my sense of touch" (Farinelli, 2009b, p. 8).

5 See Escher and Vermeulen (1989).

6 In the work of Pablo Picasso, the image of the goat is interpreted as symbolizing the overturning of conventional schemas on the part of the artist: "Pablo Picasso's predilection for goats [...] is expressed throughout his work. [...] In Spanish, 'estar como una cabra' ('to be like a goat') means 'to have lost your head' and refers to the idea of a sweet madness, the whim of someone who lets their mind wander aimlessly. Furthermore, since the Middle Ages, the reputation of the goat, and particularly the billy goat, for lasciviousness was well established; the goat was a symbol of lust and in Spanish a 'cuckhold' is a 'cabrón'. So Picasso's goat was not perhaps as innocent as it appeared" (*"Picasso. Sculptures"*, Musée National Picasso in Paris, 8 March–28 August 2016).

7 For a critical reading of *Land Art*, see Sonfist, A. (ed.) (1984), *Art in the Land: a Critical Anthology of Environmental Art*, New York: Dutton, and the documentary *Troublemakers: The Story of Land Art* (2015), director James Crump.

8 Marco Benedetti, personal communication, October 2016.

9 See Chap. 9, § 9.2.

10 Guido Scarabottolo, *Smarrimenti,* Exhibition in Palazzo Blu, Palazzo d'Arte e Cultura, Pisa, 26.05.2016.

11 Ibid.

12 Ibid.

Epilogue

Most academics prefer not to declare the particular personal interest that is hidden behind their research. In the world of science, any reference to subjective backgrounds to scientific judgement or opinion is viewed with suspicion, a threat to the validity and truth of scientific knowledge. In my own case, I have always been fascinated by the idea of getting lost, and by the labyrinth, both in its metaphorical form and as a material construction. Last but not least, in addition to my love for fairy tales such as *Hansel and Gretel* or *Alice in Wonderland* – two different but closely related ways of understanding the phenomenon of disorientation – I have personally undergone the ambiguous experience of losing my way in both the real and the metaphorical worlds.

In this sense, it is perhaps not by chance that I decided to bring a phenomenological approach to bear on my inquiry into the different manifestations of disorientation in our daily lives. In the philosophical and scientific history of phenomenology, the individual experience of both external and internal reality is a prominent focus. At the centre of systematic phenomenological investigation, we find the human subject, while recently, within a post-phenomenological framework, non-human life is also inspiring lively debate around the theme of disorientation. A phenomenological approach does not necessarily lead to the development of a theory in the strict sense of the term. I am a geographer and, therefore the obvious starting point for my investigation was the physical experience of disorientation in the real world and the concept of space. This inevitably led me to Kant as my ideal or natural point of departure, insofar as his work combines phenomenological, logical-theoretical and, above all, geographical approaches to the study of disorientation. However, while in Kant's early writings on space the subject provided the descriptive focus for the individual experience of (dis)orientation, as Kant further developed his thinking in the *Critiques*, this early phenomenological emphasis on the subject yielded to a rigid hierarchical system, which does not match my aim for a more descriptive approach.

A multi-layered phenomenon such as disorientation requires a kaleidoscopic perspective in that it cannot be easily captured using an axiomatic approach that would reduce it to a certain number of basic concepts and by following a sequence of logical steps, lead to a convincing conclusion. Indeed, it may be that the very nature of my research object has prevented me from developing a theory of disorientation. On the other hand, I believe that a systematic account of the phenomenon and a multidisciplinary overview of the state of the art in research on the theme may provide the most valid foundation for any attempt to find a reasonably satisfactory way through the labyrinthine meanders of disorientation. Hence, my study may be read as responding to the desire to better understand the experience of getting lost as a metaphor for the world we live in – a concrete condition that is part of our way of existing within this world's limits, but that also offers opportunities to overcome these limits through practices of transgression and resistance.

Marcella Schmidt di Friedberg, October 2016

References

Abbott, E. A. (1884), *Flatland: A Romance of Many Dimensions.* Oxford: Basil Blackwell.

Abe, K. (2001 [1967]), *The Ruined Map.* New York: Vintage.

Ackerman, O. E. (2009), *Getting Lost: Modes of Disorientation in Twentieth-century Literature.* Ann Arbor: Pro Quest.

Aczel, A. D. (2001), *The Riddle of the Compass: The Invention that Changed the World.* Harcourt: Houghton Mifflin.

Agamben, G. (2001), *The Open: Man and Animal.* Stanford: Stanford University Press.

Agostinis, V. (1996), *Periferie dell'anima. Labirinti, storie, voci tra rock, rap e Islam a Londra, Marsiglia e Milano.* Milano: Il Saggiatore.

Ahmed, S. (2006), *Queer Phenomenology: Orientations, Objects, Others.* Durham, NC: Duke University Press.

Ahola, M., Laaksonen, T., Sippola, K., Eeva, T., Rainio, K., and Lehikoinen, E. (2004), "Variation in Climate Warming along the Migration Route Uncouples Arrival and Breeding Dates", *Global Change Biology*, 10, pp. 1610–1617.

Alcalay, A. (1993), *After Jews And Arabs: Remaking Levantine Culture.* Minneapolis: University of Minnesota Press.

Alessandrini, A. (1993), *Pensare il bosco.* Roma: Edizioni Abete.

Alexander, M. (1993), *Fault Lines. A Memoir.* New York: The Feminist Press at CUNY.

Alexander, M. (1996), *The Shock of Arrival. Reflections on Postcolonial Experience.* Boston: South End Press.

Allen, G. L. (1999), "Spatial Abilities, Cognitive Maps, and Wayfinding: Bases for Individual Differences in Spatial Cognition and Behavior", in Golledge, R. (Ed.), *Wayfinding Behavior. Cognitive Mapping and Other Spatial Processes.* Baltimore: The John Hopkins University Press, pp. 46–80.

Allen, G. L. (2000) "Men and Women, Maps and Minds Cognitive bases of sex-related differences in reading and interpreting maps", in Ó Nuallain, S. (Ed.), *Spatial Cognition: Foundations and Applications.* Amsterdam: John Benjamins Publishing, pp. 3–18.

Allen, G. L. (Ed.) (2004), *Human spatial memory: remembering where.* Mahwah, NJ: Lawrence Erlbaum Ass.

Allison, H. (1973), *The Kant-Eberhard Controversy. An English translation together with supplementary materials and a historical-analytic introduction of Immanuel Kant's On a Discovery According to which Any New Critique of Pure Reason Has*

Been Made Superfluous by an Earlier One (Uber eine Entdeckung nach der alle neue Kritik der reinen Vernunft durch eine ältere entbehrlich gemacht werden soll). Baltimore: The John Hopkins University Press.

Almeida, I. (2004), *La boussole et la carte*. Urbino: Quattroventi.

Amery, J. (1980), *At the Mind's Limits: Contemplations by a Survivor on Auschwitz and Its Realities*. Bloomington: Indiana University Press.

Amorim, M.-A. (1999), "A Neurocognitive Approach for Human Navigation", in Golledge, R. (Ed.), *Wayfinding Behavior. Cognitive Mapping and Other Spatial Processes*. Baltimore: The John Hopkins University Press, pp. 152–167.

Anders, P. (1996), *Being There: Bemerkungen über eine cyberreale Architektur*. telepolis-Beitrag, Heise Zeitschriften Verlag, http://www.heise.de/tp, pp. 1–6.

Anders, P. (1998), "Cybrids: Integrating Cognitive and Physical Space in Architecture Convergence", *The International Journal of Research into New Media Technologies*, March, 4, 1, pp. 85–105.

Anders, P. (1999), *Envisioning Cyberspace. Designing 3D Electronic Spaces*. New York: McGraw-Hill.

Antheaume, B. and Deliry-Antheaume, E. (1999), "Cartes d'identité: les murs peints des villes d'Afrique du Sud", *Mappemonde*, 53, pp. 1–5.

Anzaldua, G. (1987), *Borderlands/La Frontera: The New Mestiza*. San Francisco: Aunt Lute Books.

Appadurai, A. (1996), *Modernity at Large: Cultural Dimension of Globalization*. Minneapolis: University of Minnesota Press.

Aragon, L. (1994 [1926]), *Paris Peasant*. Boston: Exact Change.

Ardener, S. (Ed.) (1981), *Women and Space: Ground Rules and Social Maps*. London: Croom Helm.

Ariosto, L., (1823 [1532]), *The Orlando furioso* (Transl. by Rose, W. S.). London: John Murray.

Arnold, A. E.G. F., Burles, F., and Krivoruchko, T. *et al.* (2013), "Cognitive Mapping in Humans and its Relationship to Other Orientation Skills", *Experimental Brain Research*, February, 224, 3, pp. 359–372.

Ash, J. and Simpson, P. (2014), "Geography and Post-phenomenology", *Progress in Human Geography*, 40, 1, August, pp. 48–66.

Ashihara, Y. (1989), *The Hidden Order: Tokyo through the Twentieth Century* (Trans. Riggs, L. E.). Tokyo/New York: Kodansha International.

Ashoka Britto, K. (2004), *Disorientation: France, Vietman, and the Ambivalence of Interculturality*. Hong-Kong: Hong-Kong University Press.

Attali, J. (1996), *Chemins de sagesse: Traité du labyrinth*. Paris: Fayard.

Attali, J. (1998), *Dictionnaire du XXI siècle*. Paris: Fayard.

Augé, M. (1995), *Non-places: Introduction to an Anthropology of Supermodernity*. London: Verso.

Auster, P. (1985), *City of Glass*. New York: Penguin.

Bachelard, G. (1957), *La poétique de l'espace*. Paris: Les Presses Universitaires de France.

Bailly, A. S. (1989), *I concetti della geografia umana*. Bologna: Patron Editore.

Baldacchino, G. (2005), "Islands: Objects of Representation", *Geografiska Annaler*, 87B, 4, pp. 247–251.

Baldacchino, G. (2012), "The Lure of the Island: A Spatial Analysis of Power Relations", *Journal of Marine and Islands Cultures*, 1, 2, December, pp. 55–62.

Baltrušaitis, J. (1955), *Anamorfosi o Thaumaturgus opticus*. Milano: Adelphi.

Banco, L. M., (2010), *Travel and Drugs in Twentieth-century Literature*. London: Routledge.

Barber, P. and Harper, T. (2010), *Magnificent Maps: Power, Propaganda and Art*. Chicago: University of Chicago Press.

Bard, C. (Ed.) (2004), *Le genre des territoires : féminin, masculin, neutre*. Angers: Presses de l'Université d'Angers.

Barthes, R. (1957), *Mythologies*. Paris: Le Seuil.

Barthes, R. (1982), *Empire of Signs*. New York: Hill and Wang.

Bateson, G. (1972), *Steps to an Ecology of Mind: Collected Essays in Anthropology, Psychiatry, Evolution, and Epistemology*. Chicago: University of Chicago Press.

Bateson, G. and Mead, M. (1942), *The Balinese Character: A Photographic Analysis*. Vol. II: Wilbur G. Valentine (ed.). New York: New York Academy of Sciences.

Baudelaire, C. (2008 [1863]), *Le peintre de la vie moderne*. Eloge de Constantin Guys, Collections Litteratura.Com, 30 March. http://litteratura.com.

Bauman, Z. (1998), *Globalization: The Human Consequences*. Cambridge: Polity Press.

Bauman, Z. (2003), *City of Fears, City of Hopes*. London: Goldsmiths College.

Bauman, Z. (2013), *Liquid Love: On the Frailty of Human Bond*. Hoboken, NJ: John Wiley & Sons.

Baur, R. (2009), *Orient-ierung/ation Des-/Dis-/orient-ierung/ation*. Zürich: Lars Müller Publisher.

Baur, R. (2010), *Orient-ierung/ation Des-/Dis-/orient-ierung/ation* 2. Zürich: Lars Müller Publisher.

Baynham, M. (2003), "Narratives in Space in Time: Beyond 'Backdrop' Accounts of Narrative Orientation", *Narrative Inquiry*, 13, 2, pp. 347–366.

De Beauvoir, S. (1958), *Mémoires d'une jeune fille rangée*. Paris: Gallimard. *Memoirs of a Dutiful Daughter*. (Kirkup, J. Trans.). New York: Harper and Row.

Bender, B. and Winer, M. (Eds.) (2001), *Contested Landscapes: Movement, Exile and Place*. Oxford and New York: Berg.

Benjamin, W. (1999), *The Arcades Project* (Tiedeman, R. Ed.). Cambridge, MA: Harvard University Press.

Benjamin, W. (2002), *Selected Writings: 1935–1938* (Eiland, H. and Jennings, M. W. Eds.) Vol. 3. Cambridge, MA: The Belknap Press of Harvard University Press.

Benjamin, W. (2006), *On Hashish* (Eiland, H. Ed. and Trans.). Cambridge, MA: Harvard University Press.

Benton, A. (1985), "Visuoperceptual, Visuospatial, and Visuoconstructive Disorders", in Heilman, K. M. and Valenstein, E. (Eds.), *Clinical Neuropsychology*. New York: Oxford University Press, pp. 151–185.

Berthoz, A. (1997), *Le Sens du movement*. Paris: Odile Jacob.

Bever, T. G. (1992), "The Logical and Extrinsic Sources of Modularity", in Gunnar, M. and Maratsos, M. (Eds.), *Modularity and Constraints in Language and Cognition*, Vol. 25 of *Minn Symp Child Psychol*. New York: Lawrence Earlbaum and Ass., pp. 179–212.

Biehl, J. (1991), *Finding Our Way: Rethinking Ecofeminist Politics*. Montreal: Black Rose Books.

Binet, M. A. (1894), "Reverse Illusions of Orientation", *The Psychological Review*, 1, 4, pp. 337–350.

Bittarello, M. B. (2009), "Spatial Metaphors Describing the Internet and Religious Websites: Sacred Space and Sacred Place", *Observatorio Journal*, 11, pp. 1–12.

Blumenberg, H. (1979), *Arbeit am Mythos.* Frankfurt am Main: Suhrkamp, Italian Ed. (1991), *Elaborazione del mito.* Bologna: Il Mulino.

Blumenberg, H. (1981) *Die Lesbarkeit der Welt.* Frankfurt am Main: Suhrkamp, Italian Ed. (1981), *La leggibilità del mondo.* Bologna: Il Mulino.

Blumenberg, H. (1989), *Höhlenausgänge.* Frankfurt am Main: Suhrkamp, Italian Ed. (2009), *Uscite dalla caverna.* Milano: Medusa.

Blumenberg, H. (2010), *Teoria dell'inconcettualità.* Palermo: duepunti edizioni.

Boccaccio, G. (1881 [1351]), *Stories of Boccaccio (The Decameron)* (Flameng, L. Trans.). Philadelphia: G. Barrie.

Bondi, L. (2005), "Making Connections and Thinking Through Emotions: Between Geography and Psychotherapy", School of GeoSciences, Institute of Geography Online Paper Series: GEO-004, The University of Edinburgh.

Bondi, L. and Domosh, M. (1992), "Other Figures in Other Places. On Feminism, Postmodernism and Geography", *Environment and Planning D: Society and Space,* April 1992, 10, 2, pp. 199–213.

Borges, J. L. (1964), *Labyrinths, Selected Stories & Other Writings.* New York: New Directions Publishing.

Borges, J. L. (1970), "Ibn Hakkan Al-Bokhari – Dead in His Labyrinth", *The New Yorker,* 25 April.

Borges, J. L. (1971), *The Aleph and Other Stories.* London: Penguin Classics.

Borghese, I. (Ed.) (2008), *Guido Scarabottolo. Storia di un illustratore.* Roma: Oblique Studio.

Bork, H.-R., Bork, H., Dalchov, C., Faust, B., Piorr, H.-P., and Schatz, T. (1998), *Landschafts-Entwicklung in Mitteleuropa.* Gotha: Klett-Perthes.

Boudjedra, R. (1975), *Topographie idéale pour une aggression caractérisée.* Paris: Denoêl.

Boyle, P., Halfacree, K., and Robinson, V. (2014), *Exploring Contemporary Migration.* London: Routledge.

Braidotti, R. (1996), *Madri, mostri e macchine.* Manifestolibri: Roma.

Braidotti, R. (2002), *Nuovi soggetti nomadi.* Sossella: Roma.

Braidotti, R. (2003), "Becoming Woman: or Sexual Difference Revisited", *Theory, Culture and Society,* 20, 3, pp. 43–64.

Braidotti, R. (2013), *The Posthuman.* Cambridge: Polity Press.

Bramwell, A. (1989), *Ecology in the 20th Cantury: A History.* New Haven, CT: Yale University Press.

Brandt, W. (1980), *North-South: A Programme for Survival: Report at the Independent Commission on International Development Issues.* Cambridge, MA: MIT Press.

Braudel, F. (1995 [1966]), *The Mediterranean and the Mediterranean World in the Age of Philip II,* Volume 1. Berkley: University of California Press.

Brayton, D. (2012), *Shakespeare's Ocean: An Ecocritical Exploration.* Charlottesville: University of Virginia Press.

Brenner, N. and Keil, R. (Eds.) (2006), *The Global Cities Reader.* Routledge: New York.

Breton, A. (1969 [1924]), *Manifestoes of Surrealism* (Seaver, R. and Lane, H. R. Trans.). Ann Arbor Paperbacks: University of Michigan Press.

Breton, A. (1973), *Entretiens.* Paris: Gallimard.

Briginshaw, V. (2009), *Dance, Space and Subjectivity.* New York: Palgrave MacMillan.

Brown, C. H. (1983), "Where Do Cardinal Directions Come From?", *Anthropological Linguistics,* 25, pp. 121–161.

Brunswik, E. (1947), *Systematic and Representative Design of Psychological Experiments.* Berkeley: University of California Press.

Bryant, K. J. (1982), "Personality Correlates of Sense of Direction and Geographical Orientation", *Journal of Personality and Social Psychology*, 43, pp. 1318–1324.

Burbules, N. C. (1997), "Aporia: Webs, Passages, Getting Lost, and Learning to Go On", *Philosophy of Education Yearbook*, http://www.ed.uiuc.edu/EPS/PES-yearbook/97_docs/burbules.html.

Burroughs, W. and Ginsberg, A. (2006), *The Yage Letters* (Harris, O. Ed.). San Francisco: City Lights Books.

Butler, J. (1993), *Bodies that Matter: On the Discursive Limits of "Sex"*. London: Routledge.

Butler, J. (1999), *Gender Trouble*, 2nd edition. London: Routledge.

Butler, J. (2015), *Senses of the Subject*. New York: Fordham University Press.

Buttimer, A. (1976), "Grasping the Dynamism of the Lifeworld", *Annals of the Association of American Geographers*, 66, 2, pp. 277–292.

Cacciari, M. (2005), "'Lichtung': intorno a Heidegger e Maria Zambrano", in Petterlini, A., Brianese, G. and Goggi, G. (a cura di), *Le parole dell'Essere. Per Emanuele Severino*. Milano: Bruno Mondadori, pp. 123–130.

Cadet, B. (2010), "Orientation, Disorientation, and Context: The Cognitive Perspective", in Baur, R., *Orient-ierung/ation Des-/Dis-/orient-ierung/ation* 2. Zürich: Lars Müller Publisher, pp. 205–218.

Callicott, J. B. and Nelson, M. P. (Eds.) (1998), *The Great New Wilderness Debate*. Athens, GA: University of Georgia Press.

Calvino, I. (1974), *Invisible Cities* (Weaver, W. Trans.). Orlando: Harcourt.

Campbell, A. (1999), "Staying Alive: Evolution, Culture, and Women's Intrasexual Aggression", *Behavioral and Brain Science*, 22, pp. 203–252.

Capel, H. (2003), "Una mirada histórica sobre los estudios de redes de ciudades y sistemas urbanos", *GeoTrópico*, 1, 1, GeoLat, Bogotà, pp. 30–65, http://www.geotropico.org/1_1_Capel.html.

Capel, H. (2004), "El futuro de las ciudades. Una propuesta de manifesto", Biblio 3W, *Revista Bibliográfica de Geografía y Ciencias Sociales*, Universidad de Barcelona, Vol. IX, n° 551, 10 de diciembre, http://www.ub.es/geocrit/b3w-551.htm.

Careri, F. (2001), *Constant. New Babylon, una città nomade*. Torino: Testo & Immagine.

Careri, F. (2006), *Walkscapes. Camminare come pratica estetica*. Torino: Einaudi.

Carlson, J. B. (1975), "Settlement and Cultural Development at Chalcatzingo", *Science*, New Series, 192, 4245, pp. 1203–1210.

Carroll, J. B. (Ed.) (1956), *Language, Thought, and Reality: Selected Writings of Benjamin Lee Whorf*. Cambridge, MA: MIT Press.

Carroll, L. (1985 [1876]), *La caccia allo Snualo. The Hunting of the Snark*. Pordenone: Studio Tesi.

Casamajor, J. (1927), "Le Mysterieux Sens de l'Espace", *Revue Scientiphique*, 65, 18, pp. 554–565.

Casselman, A. (2008), "The Earth Has More Than One North Pole: The North Pole is More than Just the Top of the Planet", *Scientific American*, Thursday, 28 February.

Castells, M. (1996), *The Rise of Network Society*. Cambridge: Blackwell.

Castells, M. (2006), "Cities, the Information Society and the Global Economy", in Brenner, N. and Keil, R. (Eds.), *The Global Cities Reader*. New York: Routledge, pp. 135–137.

Casti, E. (1998), *L'ordine del mondo e la sua rappresentazione*. Milano: Unicopli.

Cawood, J. (1979), "The Magnetic Crusade: Science and Politics in Early Victorian Britain", *Isis*, 70, pp. 493–519.

De Certeau, M. (1984), *The Practice of Everyday Life* (Rendall, S. F. Trans.). Berkeley: University of California Press.

Chaffrai, S. (2005), "Corps, territoire et paysage à travers les images et les textes viatiques en Nouvelle-France (1701–1756)", *Revue d'histoire de l'Amérique française*, 59, 1–2, pp. 7–52.

Chandler, S. C. (1891), "On the Variation of Latitude, II", *The Astronomical Journal*, 11, 249, pp. 65–70.

De Charlevoix, P. F. X. (1744), *Histoire et Description Generale de la Nouvelle France avec Le Journal Historique d'un Voyage fait par ordre du Roi dans l'Amérique Septentrionnale.* Paris: Didot, 3 vol.

Chomsky, N. (2016), *What Kind of CREATURES Are We?* New York: Columbia University Press.

Civati, G. (2005), *"C'è troppo nulla qui". Appunti per una storia dell'idea di foresta.* Milano: Cuem.

Clark, A. (1997), *Being There: Putting Brain, Body and World Together.* Cambridge, MA: MIT Press.

Clark, T. K. *et al.* (2010), "Analysis of Human Spatial Perception during Lunar Landing", *Aerospace Conference, 2010 IEEE*, pp. 1–13.

Clynes, M. and Kline, N. S. (1960), "Cyborgs and Space", *Astronautics*, 14, September Issue, pp. 26–27, pp. 74–75.

Cohen, J. E. (2007), "Cyberspace as/and Space", *Columbia Law Review*, 107, 1, pp. 210–256.

Cohen, L. A. (1965), "Hallucinogenic Spatial Disorientation", *International Journal of Neuropsychiatry*, 1, 4, pp. 347–351.

Collard, A.-S. (2006), "La métaphore dans l'hypermédia comme médiateur des contenus", in Ferjani, R. (Ed.), *Technologies de l'Information et de la Communication. Discours, représentations et pratiques.* La Manouba (Tunis): Institut de la Presse et des Sciences de l'information, pp. 85–103.

Conrad, J. (2006 [1899]), *Heart of Darkness.* Minneapolis: Filiquarian Publishing, LLC.

Cooke, J. H. (1883), "The Shipwreck of Sir Cloudesley Shovell on the Scilly Islands in 1707", From Original and Contemporary Documents Hitherto Unpublished, Read at a Meeting of the Society of Antiquaries, London, 1 February.

Corbella, D. (1996), "El mito de San Borodòn: entre realidad y fabùla", in Fernando Carmona Fernández and Perez A. Martinez (Eds.), *Libros de viaje: actas de las Jornadas sobre los Libros de Viaje en el Mundo Románico*, celebradas en Murcia del 27 al 30 de noviembre de 1995, Universidad de Murcia, pp. 127–136.

Cornell, E. H., Sorenson, A. and Mio, T. (2003), "Human Sense of Direction and Wayfinding", *Annals of the Association of American Geographers*, 93, 2, pp. 399–425.

Cornetz, V. (1909), "Observations sur le sens de l'orientation chez l'homme", *La Revue des Idées*, 67, pp. 60–65.

Coverly, M. (2006), *Psychogeography.* London: Pocket Essentials.

Crang, M. (2001), "Rhythms of the City: Temporalised Space and Motion", in May, J. and Thrift, N. (Eds.), *Timespace: Geographies of Temporality.* London: Routledge, pp. 187–207.

Cresswell, T. (1996), *In Place/out of Place: Geography, Ideology, and Transgression.* Minneapolis: University of Minnesota Press.

Cresswell, T. (2006), *On the Move: Mobility in the Modern Western World.* London: Routledge.

Cresswell, T. (2009), "Place", in Thrift, N. and Kitchen, R. (Eds.), *International Encyclopedia of Human Geography*. Oxford: Elsevier, Vol. 8, pp. 169–177.

Cresswell, T. (2014a), *Place: An Introduction*. Hoboken, NJ: John Wiley & Sons.

Cresswell, T. (2014b) "Place", in Cloke, P. J., Crang, P. and Goodwin, M. (Eds.), *Introducing Human Geographies*. London: Routledge, pp. 249–261.

Cronon, W. (Ed.) (1995), *Uncommon Ground: Rethinking the Human Place in Nature*. New York: W. W. Norton & Co.

Czerwinski, M., Tan, D. S., and Robertson, G. G. (2002), "Women Take a Wider View", *CHI 2002*, 4, 1, Minneapolis, April 2002.

Dabbs, J. M., Chang, E. L., Strong, L. A., and Milun, R. (1998), "Spatial Ability, Navigation Strategy, and Geographic Knowledge Among Men and Women", *Evolution and Human Behaviour*, 19, 2, pp. 89–98.

Damasio, A. (2012), *Self comes to Mind: Constructing the Conscious Brain*. London: Vintage.

Dante, A. (1830 [1472]), *La Divina Commedia di Dante Alighieri col comento del P. Baldassarre Lombardi*. Firenze: Leonardo Ciardetti.

Darwin, C. (1859), *On the Origin of Species by Means of Natural Selection, or the Preservation of Favoured Races in the Struggle for Life*. London: John Murray.

Darwin, C. (1871), *The Descent of Man, and Selection in Relation to Sex*. London: John Murray.

Darwin, C. (1873), "Origin of Certain Instincts", *Nature*, 3, 179, pp. 417–418.

Darwin, C. (1883), "Posthumous Essay on Instinct", in Romanes, G. J., *Mental Evolution in Animals*, London: Kegan Paul Trench & Co., pp. 355–359.

Davidson, J. (2000), "A Phenomenology of Fear: Merleau-Ponty and Agoraphobic Life worlds", *Sociology of Health and Illness*, 22, pp. 640–660.

Davidson, J. (2003), *Phobic Geographies: The Phenomenology and Spatiality of Identity*. Aldershot: Ashgate.

Davidson, J., Smith, M., and Bondi, L. (Eds.) (2005), *Emotional Geographies*. Aldershot: Ashgate.

Davidson, P. (2016), *The Idea of North*. London: Reaktion Books.

Davis, G. and Dwyer, C. (2007), "Qualitative Methods: Are You Enchanted or Are You Alienated?", *Progress in Human Geography*, 31, 2, pp. 257–266.

Debord, G. (1955a), "Introduction à une critique de la géographie urbain", *Les lèvres nues*, 6, Bruxelles, pp. 11–15.

Debord, G. (1955b), "Panorama intelligent de l'avant-guarde à la fin de 1955", *Potlach*, 24, pp. 1–6.

Debord, G. (1956), "Theorie de la derive", *Les lèvres nues*, 9, pp. 19–23.

Debord, G. (1994 [1967]), *The Society of the Spectacle*. New York: Zone Books.

Deleuze, G. (2004), *Desert Islands and Other Texts 1953–1974*. Los Angeles: Semiotext.

Deleuze, G. and Guattari, F. (1987), *A Thousand Plateaus: Capitalism and Schizophrenia* (Massumi, B. Trans.). Minneapolis: University of Minnesota Press. (1980), *Capitalisme et schizophrénie*, Tome 2, *Mille Plateaux*. Paris: Éditions de Minuit.

Deleuze, G., and Guattari, F. (1991), "Qu'est-ce que la philosophie?", in *Philosophie, science et art*. Paris: Les Editions de Minuit, pp. 189–190.

Deleuze, G., and Guattari, F. (1994), *What Is Philosophy?* New York: Columbia University Press.

Delgado, C. F. and Muñoz, J. E. (1997), *Everynight Life: Culture and Dance in Latin/o America*. Durham, NC: Duke University Press.

dell'Agnese, E. (2015), "Svolte sbagliate? Stereotipi sociali e paesaggi della paura nella tradizione cinematografica del backwoods horror", in *Oltre il libro: nuove forme di narratività sugli/dagli USA*, Ácoma n. 9, Autunno-Inverno, pp. 56–68.

Dematteis, G. (1985), *Le metafore della terra*. Milano: Feltrinelli.

Derrida, J. (1985 [1980]), "To Speculate - on 'Freud'", in Blonsky, M. (Ed.), *On Signs*. Baltimore: The John Hopkins University Press, pp. 236–258.

Derrida, J. (2012), "The Rhetoric of Drugs" (Israel, M. Trans.), in Alexander, A. and Roberts, M. S. (Eds.), *High Culture: Reflections on Addiction and Modernity*. Albany, NY: SUNY Press, pp. 19–44.

Descartes, R. (2009 [1637]), *Discourse on the Method: Of Rightly Conducting One's Reason and of Seeking Truth in the Sciences*. Auckland: The Floating Press.

Devlin, A. S. and Bernstein, J. (1995), "Interactive Wayfinding: Use of Cues by Men and Women", *Journal of Environmental Psychology*, 15, pp. 22–38.

Dingle, H. (1996), *Migration: The Biology of Life on the Move*. Oxford: Oxford University Press.

Dodge, M. and Kitchin, R. (2001), *Atlas of Cyberspace*. Harlow: Pearson Education.

Dudchenko, P. (2010), *Why People Get Lost. The Psychology and Neuroscience of Spatial Cognition*. Oxford: Oxford University Press.

Dünne, J. and Günzel, S. (Eds.) (2006), *Raumtheorie. Grundlagentexte aus Philosophie und Kulturwissenschaften*. Frankfurt am Main: Suhrkamp Verlag.

Eals, M. and Silverman, I. (1994), "The Hunter-gather Theory of Spatial Sex Differences: Proximate Factors Mediating the Female Advantage in Recall of Object Arrays", *Ethnology and Sociobiology*, 15, pp. 95–105.

Eco, U. (1984), "Prefazione", in Santarcangeli, P. *Il libro dei labirinti*. Milano: Frassinelli, pp. XI–XV.

Eco, U. (2014), *From the Tree to the Labyrinth: Historical Studies on the Sign and Interpretation* (Oldcorn, A. Trans.). Cambridge, MA: Harvard University Press. Italian Ed. (2007), *Dall'albero al labirinto. Studi storici sul segno e l'interpretazione*. Milano: Bompani.

Ecuyer-Dab, I., Robert, M. (2004), "Have Sex Differences in Spatial Ability Evolved from Male Competition for Mating and Female Concern for Survival?", *Cognition*, 91, 3, April, pp. 221–257.

Eichberg, H. (1998), "The Labyrinth of the City – Fractal Movement and Identity", in Pedersen, K. and Viken, A. (Eds.), *Nature and Identity: Essays on the Culture of Nature*. Kristiansand: Høyskoleforlaget, pp. 265–301.

Von Eichendorff, J. (1984 [1815]), "Waldgespräch", in Segebrecht, W. (Ed.), *Gedichte und Interpretationen. Klassik und Romantik*. Stuttgart: Reclam, vol. 3, p. 306.

Eitel, E. J. (2012 [1873]), *What is Feng-Shui? The Classic Nineteenth Century Interpretation*. Mineola, NY: Dover Publications.

Ekman, K. (2008), *Der Wald. Eine literarische Wanderung*. Munchen: Piper.

Elden, S. (2009), "Space I", in Thrift, N. and Kitchin, R. (Eds.), *International Encyclopedia of Human Geography*. Amsterdam: Elsevier, pp. 262–267.

Elden, S. (2011), "Reintroducing Kant's Geography", in Elden, S. and Mendieta, E. (Eds.), *Reading Kant's Geography*. Albany: State University of New York Press, pp. 1–15.

Elden, S. and Mendieta, E. (Eds.) (2011), *Reading Kant's Geography*. Albany: State University of New York Press.

Eliade, M. (1976 [1968]), *Miti, sogni e misteri*. Milano: Rusconi Eliminated. *Myths, Dreams and Mysteries*. New York: Harper & Row.

Ellard, C. (2009), *You Are Here: Why We Can Find Our Way to the Moon but Get Lost in the Mall*. New York: Doubleday.
Emerson, R. W. (1983 [1836]), *Nature: Addresses and Lectures*. New York: Literary Classics of the U.S., pp. 6–49.
Emlen, S. T. (1975), "The Stellar-orientation System of a Migratory Bird", *Sci. Am.*, 233, pp. 102–111.
Emlen, S. T. and J. T. Emlen (1966), "A Technique for Recording Migratory Orientation of Captive Birds", *The Auk*, 83, pp. 361–367.
Engeli, M. (2000), *Digital Stories: The Poetics of Communication*. Basl: Birkhäuser.
Engels, S., Schneider, N.-L., Lefeldt, N., Hein, C. M., Zapka, M., *et al.* (2014), "Anthropogenic Electromagnetic Noise Disrupts Magnetic Compass Orientation in a Migratory Bird", *Nature*, 509, 7500, pp. 353–356.
Escher, Maurits Cornelis and Vermeulen, J. W. (1989), *Escher on Escher: Exploring the Infinite*. New York: Abrams.
Evernden, N. (1985), *The Natural Alien: Humankind and Environment*. Toronto: University of Toronto Press.
Everts, T. C. (1871), "Thirty-seven Days of Peril", *Scribner's Monthly*, November, 3, pp. 1–17.
Fanelli, M. C. (1997), *Labirinti. Storia, geografia e interpretazione di un simbolo millenario*. Rimini: Il Cerchio.
Fanon, F. (1963 [1961]), *The Wretched of the Earth* (Preface by Sartre, J.-P., Farrington, C. Trans.). New York: Grove Press.
Fanon, F. (2008 [1952]), *Black Skin, White Masks* (Markmann, C. L. Trans.). London: Pluto Press.
FAO (2012), *State of the World's Forests 2012*. Rome: Food and Agriculture Organization of the United Nations.
Farinelli, F. (2002), "The Geography of Zero", in Picone, M. (a cura di), *Bodies and Space. Gunnar's Travels*. Palermo: Laboratorio Geografico, pp. 79–88.
Farinelli, F. (2003), *Geografia. Un'introduzione ai modelli del mondo*. Torino: Einaudi.
Farinelli, F. (2009a), *La crisi della ragione cartografica*. Torino: Einaudi.
Farinelli, F. (2009b), "Lo spazio, il luogo e la crisi della ragione cartografica", Spaziocinema, Laboratorio Probabile Bellamy, http://www.laboratorioprobabile.it/allegati/farinelli.pdf.
Featherstone, M. (1998), "The Flâneur, the City and Virtual Public Life", *Urban Studies*, 35(5–6), pp. 909–925.
Feddersen, E. and Lüdtke, I. (Eds.) (2014), *Lost in Space*. Basel: Birkhäuser.
De Fina, A. (2003), "Crossing Borders: Time, Space, and Disorientation in Narrative", *Narrative Inquiry*, 13, 2, pp. 367–391.
Finney, B. (1979), *Hokulea: The Way to Tahiti*. New York: Dodd, Mead and Company.
Finney, B. (1994), *Voyage of Rediscovery: A Cultural Odyssey through Polynesia*. Berkeley: University of California Press.
Finney, B. (1997), "Exploring and Colonizing the Pacific by Logic and Computer", *Pacific Studies*, 20, 2.
Finney, B. (1998), "Nautical Cartography and Traditional Navigation in Oceania", in Woodward, D. and Lewis, G. M. (Eds.), *The History of Cartography*, Vol. 2, Book 3 *Cartography in Prehistoric, Ancient, and Medieval Europe and the Mediterranean*. Chicago: University of Chicago Press, pp. 443–492.
Finney, B. (2003), *Sailing in the Wake of the Ancestors: Reviving Polynesian Voyaging*. Honolulu: Bishop Museum Press.

Fisher, A. (2006), *The Amazing Book of Mazes.* London: Thames & Hudson.

Fitzpatrick, S. M. (2007), "Archaeology's Contribution to Island Studies", *Island Studies Journal*, 2, 1, pp. 77–100.

Forde, H. (1873), "Sense of Direction", *Nature*, 7, April 17, pp. 463–464.

Fortescue, M. (1988), *Eskimo Orientation Systems.* Copenhagen: Museum Tusculanum Press.

Foucault, M. (1972), *Folie et déraison. Histoire de la folie à l'âge classique.* Paris: Gallimard.

Foucault, M. (1984), "Space, Knowledge and Power", Interview with Paul Rabinov, in Rabinow, P. (Ed.), *The Foucault Reader.* New York: Pantheon, pp. 239–256.

Foucault, M. (1986), "Of Other Spaces", *Diacritics*, 1, 16, pp. 22–27.

Foucault, M. (1997 [1976–1988]), *Dits et Ecrits*, tome 2. Paris: Gallimard.

Freud, S. (1919), "The 'Uncanny'", *The Standard Edition of the Complete Psychological Works of Sigmund Freud.* (Transl. under the General Editorship of Strachey, J., in collaboration with Anna Freud), 24 volumes. Vancouver: Vintage, 1999. Volume XVII (1917–1919): "An Infantile Neurosis and Other Works", pp. 217–256.

Gaard, G. (1997), 'Toward a Queer Ecofeminism', *Hypatia*, 12, 1, pp. 114–137.

Galea, L. A. M. and Kimura, D. (1993), "Sex Differences in Route Learning", *Personality and Individual Differences*, 14, pp. 53–65.

Galton, F. (1853), *The Narrative of an Explorer in Tropical South Africa.* London: John Murray.

Garfield, S. (2012), *On the Map: A Mind-Expanding Exploration of the Way the World Looks.* London: Profile Books.

Gaulin, S. J. C. and Fitzgerald, R. W. (1989), "Sex Differences in Spatial Ability: An Evolutionary Hypothesis and Test", *American Naturalist*, 127, pp. 74–88.

Gauthreaux, S. A. (Ed.) (1980), *Animal Migration, Orientation and Navigation.* New York: Academic Press.

Geary, D. C. (1995), "Sexual Selection and Sex-differences in Spatial Cognition", *Learning and Individual Differences*, 7, pp. 289–301.

Geertz, C. (1973), *Interpretation of Cultures: Basic Essays.* New York: Basic Books.

Gibson, W. (1982), "Burning Chrome", *Omni*, 4, 10, p. 72.

Gibson, W. (1984), *Neuromancer.* New York: Ace Books.

Gibson, W. (1988), *Mona Lisa Overdrive.* New York: Bantam Books.

Gilmartin, P. P. and Patton, J. C. (1984), "Comparing the Sexes on Spatial Abilities: Map-Use Skills", *Annals of the Association of American Geographers*, 74, 4, pp. 605–619.

Gladwin, T. (1970), *East is a Big Bird: Navigation and Logic on Puluwat Atoll.* Cambridge, MA: Harvard University Press.

Gluck, M. (1991), *Making Sense of Human Wayfinding: Review of Cognitive and Linguistic Knowledge for Personal Navigation with a New Research Direction.* New York: Springer.

De Gobineau, A. (1967 [1854]), *Essai sur l'inegalité des race humaines*, t. II. Paris: Éditions Pierre Belfond.

Golledge, R. (Ed.) (1999), *Wayfinding Behavior: Cognitive Mapping and Other Spatial Processes.* Baltimore: The John Hopkins University Press.

Golledge, R., Dougherty, V., and Bell, S. (1995), "Acquiring Spatial Knowledge: Survey Versus Route-Based Knowledge in Unfamiliar Environments", *Annals of the Association of American Geographers*, 85, pp. 134–158.

Gottschild, B. D. (2003), *Black Dancing Body: A Geography from Coon to Cool*. New York: Palgrave Macmillan.

Gould, P. and Olsson, A. (1982), *A Search for Common Ground*. London: Pion Limited.

Graves, R. (1948), *The White Goddess: A Historical Grammar of Poetic Myth*. London: Faber & Faber.

Greene, T. M. (2001), "Labyrinth Dances in the French and English Renaissance", *Renaissance Quarterly*, 54, 4.2, February, pp. 1403–1466.

Gregson, N. and Rose, G. (2000), "Taking Butler Elsewhere: Performance, Spatialities and Subjectivities", *Environment and Planning D: Society and Space*, 18, 4, pp. 433–452.

Griffith, B., Scott, J. M., Carpenter, J. W., and Reed, C. (1989), "Translocation as a Species Conservation Tool: Status and Strategy", *Science*, 245, pp. 477–482.

Günzel, S. (Ed.) (2009), *Raum-wissenschaften*. Frankfurt am Main: Suhrkamp Verlag.

Hagstrum, J. T. (2013), "Atmospheric Propagation Modeling Indicates Homing Pigeons Use Loft-specific Infrasonic 'Map' Cues", *J. Exp. Biol.*, 216, pp. 687–699.

Halpern, D. F., Benbow, C. P., Geary, D. C., Gur, R. C., Hyde, J. S., and Gernsbacher, M. A. (2007), "The Science of Sex Differences in Science and Mathematics", *Psychological Science in the Public Interest*, 8, 1, pp. 1–51.

Hancock, C. (2004), "L'idéologie du territoire en géographie : incursions féminines dans une discipline masculiniste", in Bard, C. (Ed.), *Le genre des territoires : masculin, féminin, neutre*. Angers: PUA, pp. 167–176.

Haraway, D. (1991), "A Cyborg Manifesto: Science, Technology, and Socialist-Feminism in the Late Twentieth Century", in *Simians, Cyborgs and Women: The Reinvention of Nature*. New York: Routledge, pp. 149–181.

Hardy, G. (1933), *Géographie et colonization*. Paris: Gallimard.

Harley, J. B. (1987), "The Map and the Development of the History of Cartography" in Harley, J. B. and Woodward, D. (Eds.), *The History of Cartography*. Chicago: University of Chicago Press, vol. 1, pp. 1–42.

Harley, J. B. (1989), "Deconstructing the Map", *Cartographica*, 26, 2, Spring, pp. 1–20.

Harris, O. (2003), *William Burroughs and the Secret of Fascination*. Carbondale and Edwardsville: Southern Illinois University Press.

Harris, P. (1996), *HYPER-LEX: A Technographical Dictionary*, http://www.electronicbookreview.com/thread/criticalecologies/lexical. Accessed 10 April 2017.

Harrison, R. P. (1992), *Forests: The Shadow of Civilization*. Chicago: University of Chicago Press.

Hart, R. A. (1979), *Children's Experience of Place*. New York: Irvington.

Harvey, D. (2007 [1969]), *Explanation in Geography*, London: Edward Arnold.

Harvey, D. (2009), *Cosmopolitanism and the Geographies of Freedom*. New York: Columbia University Press.

Harvey, D. (2012), *Rebel Cities: From the Right to the City to the Urban Revolution*. London: Verso.

Hassabis, D., Kumaran, D., Vann, S. D., and Maguire, E. A. (2007), "Patients with Hippocampal Amnesia Cannot Imagine New Experiences", *PNAS*, 104, 5, pp. 1726–1731.

Hay, P. A. (2006), "Phenomenology of Islands", *Islands Studies Journal*, 1, 1, pp. 19–42.

Hayes, D. M. (2003), *Body and Sacred Place in Medieval Europe, 1100–1389: Interpreting the Case of Chartres Cathedral*. London: Routledge.

Hedges, I. (1983), *Languages of Revolt: Dada and Surrealist Literature and Film*. Durham, NC: Duke University Press.

Heidegger, M. (1971 [1951]), "Building Dwelling Thinking", in *Poetry, Language, Thought*, Hofstadter, A. Trans. New York: Harper Colophon Books, pp. 145–161.

Heidegger, M. (2002 [1950]), *Off the Beaten Track* (Young, J. and Hayn, K. Eds. and Transl.). Cambridge: Cambridge University Press.

Heine, B. (1997), *Cognitive Foundations of Grammar*. New York, Oxford: Oxford University Press.

Hemmings, C. (2002), *Bisexual Spaces: A Geography of Sexuality and Gender*. London: Routledge.

Herbert, A. D. (1970), "Spatial Disorientation in Birds", *Wilson Bull.*, 82, pp. 400–419.

Hertz, R. (1960 [1909]), *The Pre-Eminence of the Right Hand. A Study in Religious Polarity*. Translated by Needham, R. and Needham, C. London: Cohen & West.

Highmore, B. (2005), *Cityscapes: Cultural Readings in the Material and Symbolic City*. Basingstoke and New York: Palgrave Macmillan.

Hildburgh, W. L. (1945), "The Place of Confusion and Indeterminability in Mazes and Maze-Dances", *Folklore*, 56, 1, pp. 188–192.

Hill, K. A. (1998), *Lost Person Behavior*. Ottawa: National SAR Secretariat.

Hirsch, E. D. (1976), *The Aims of Interpretation*. Chicago: University of Chicago Press.

Hirsch, E. G. and Benzinger, I. (1906), *Winds*, Jewish Encyclopedia, jewishencyclopedia.com.

Hocke, G. R. (1957), *Die Welt als Labyrinth*. München: Rowohlt.

Holloway, J. B. (1981), "Introduction" to *Il Tesoretto* by Latini, Brunetto H. (Ed. and Transl.). New York: Grand Publishing, Inc.

Hornbacher, A. (2010), "Localization or Imagination? (Dis)orientation in Global Modernity", in Baur, R., *Orient-ierung/ation Des-/Dis-/orient-ierung/ation* 2. Zürich: Lars Müller Publisher, pp. 95–98.

Hubbard, P. (2008), *Here, There, Everywhere: The Ubiquitous Geographies of Heteronormativity*, https://dspace.lboro,ac.uk. Accessed 10 April2017.

Humphrey, C. (2001), "Contested Landscapes in Innert Mongolia: Walls and Cairns", in Bender, B. and Winer, M. (Eds.), *Contested Landscapes: Movement, Exile and Place*. Oxford and New York: Berg, pp. 55–68.

Hunt, G., Milhet, M., and Bergeron, H. (Eds.) (2011), *Drugs and Culture: Knowledge, Consumption and Policy*. Farnham: Ashgate.

Hutchins, E. (1983), "Understanding Micronesian Navigation", in Gentner, D. and Stevens, A. (Eds.), *Mental Models*. Hillsdale, NJ: Lawrence Erlbaum, pp. 191–225.

Huth, J. E. (2013), *The Lost Art of Finding Our Way*. Cambridge, MA: Belknap Press.

Huxley, A. (1954), *The Doors of Perception*. New York: Harper & Brothers.

Huxley, A. (1956), *Heaven and Hell*. London: Chatto & Windus.

Iaria, G., Bogod, N., Fox, C. J., and Barton, J. S. (2008), "Developmental Topographical Disorientation: Case One", *Neuropsychologia*, 47, pp. 30–40.

Ieranò, G. (2007), *Il mito di Arianna. Da Omero a Borges*. Roma: Carocci.

Ihde, D. (2009), "Postphenomenology and Technoscience", The Peking University Lectures. New York: State University of New York Press.

Ingold, T. (2007), *Lines: A Brief History*. London: Routledge.

Irigaray, L. C. (1985), *This Sex Which is Not One* (Porter, C. Trans.), Burke, Ithaca, New York: Cornell University Press.

Jaccard, P. (1932), *Le sens de la direction et l'orientation lointaine chez l'homme*. Paris: Payot.

James, W. (1907), *Pragmatism (A New Name for Some Old Ways of Thinking)*. New York: Longmans, Green, and Co.

James, W. (1950), *The Principles of Psychology.* New York: Dover Publications.

Jenning, K. (2011), *Maphead: Charting the Wide, Weird World of Geography Wonks.* New York: Simon & Schuster.

John, G. E. and Metzo, C. R. (2008), "Yellowstone Embodied: Truman Everts's 'Thirty-Seven Days of Peril'", *Gender, Place, and Culture,* 15, 3, June, pp. 221–242.

Johnson, T. (2009), "Transgressing the Territories of Dance: Spatiality and Salsa in North Carolina", paper presented at the 2009 Congress of the Latin American Studies Association, Rio de Janeiro, Brazil, June 11–14.

Jones, O. and Cloke, P. (2002), *Tree Cultures. The Place of Trees and Trees in Their Place.* Oxford: Berg.

Jonsson, E. (2002), *Inner Navigation: Why We Get Lost and How We Find Our Way.* New York: Scribner.

Juhel, P. (2013), *Histoire de la boussole: L'aventure de l'aiguille.* Versailles: Editions Quae.

Kafka, F. (1992), *Das Ehepaar und andere Schrifte aus dem Nachlaß.* Frankfurt: Fischer.

Kafka, F. (2016), *Gesammelte Werke.* Duesseldorf: Null Papier Verlag.

Kajmaku, E. (2016), "The Labyrinth of the Third Millennium", *Lab2.0 Learning Architecture & Building,* 14, 12-2015/1-2016, p. 33.

Kant, I. (1983), *Anthropologie in pragmatischer Sicht.* Stuttgart.

Kant, I. (1992 [1768]), "Concerning the Ultimate Ground of the Differentiation of Directions in Space", in Walford, D. and Meerbote, R. (Trans. and Eds.), *Theoretical Phylosophy, 1755–1770,* The Cambridge Edition of the Works of Immanuel Kant (Guyer, P. and Wood, A. W., Eds.). Cambridge: Cambridge University Press, pp. 361–372.

Kant, I. (1992 [1770]), "On Space", §15, "On the form and principles of the sensible and the intelligible world, Inaugural Dissertation", in Walford, D. and Meerbote, R. (Trans. and Eds.), *Theoretical Phylosophy, 1755–1770,* The Cambridge Edition of the Works of Immanuel Kant (Guyer, P., Wood, A. W., Eds.). Cambridge: Cambridge University Press, pp. 373–400.

Kant, I. (1996 [1786]), "What Does It Mean To Orient Oneself in Thinking?", in (Wood, A. W. Trans.), *Religion and Rational Theology,* The Cambridge Edition of the Works of Immanuel Kant (Guyer, P. and Wood, A. W. Eds.). Cambridge: Cambridge University Press, pp. 1–18.

Kant, I. (1998 [1781]), *Critique of Pure Reason.* Guyer, P. and Wood, A. W. Trans. and Ed., The Cambridge Edition of the Works of Immanuel Kant. Cambridge: Cambridge University Press.

Kant, I. (1998), "On the ground of the distinction of all objects in general into phenomena and noumena" (as in the second edition), pp. 354–366 (in or. pp. 152–176).

Kant, I. (2001 [1786]), "What Does it Mean to Orient Oneself in Thinking", in *Religion and Rational Theology,* The Cambridge Edition of the Works of Immanuel Kant, Guyer, P. and Wood, A. W. Trans. and Ed. Cambridge: Cambridge University Press.

Kant, I. (2006 [1798]), *Anthropology from a Pragmatic Point of View* (Louden, R. B. Trans. and Ed.). Cambridge: Cambridge University Press.

Kerényi, C. (1983), *Nel labirinto.* Milano: Bollati Boringhieri.

Kern, H. (1982), *Labyrinthe. Erschinungsformen und Deutungen. 5000 Jahre Gegenwart eines Urbild.* München: Prestel Verlag.

Kimura, D. (1998), "Sex, Sexual Orientation and Sex Hormones Influence Human Cognitive Functions", in Squire, L. and Kosslyn, S. M., *Findings and Current Opinion in Cognitive Neuroscience.* Cambridge, MA: MIT Press, pp. 155–160.

Kitchin, R. and Kneale, J. (2002), *Lost in Space. Geographies of Science Fiction.* London: Continuum.

Klaproth, J. (1834), *Lettre à M. le baron de Humboldt sur l'invention de la boussole.* Paris: Librairie orientale Dondey-Dupré.

Knight, J. (1997), "A Tale of Two Forests: Reforestation Discourse in Japan and Beyond", *Journal of the Royal Anthropological Institute*, 3, 4, pp. 711–728.

Knight, J. (2000), "From Timber to Tourism: Recommoditizing the Japanese Forest", *Development and Change*, 31, pp. 341–359.

Koester, R. J. (2008), *Lost Person Behavior: A Search and Rescue Guide on Where to Look – for Land, Air and Water.* Charlottesville: dbS Productions.

Komensky, J. A. (1901 [1631]), *The Labyrinth of the World and the Paradise of the Heart.* New York: E. P. Dutton & Co. Silvio Berlusconi Editore.

Konijnendijk, C. C. (2008), *The Forest and the City: The Cultural Landscape of Urban Woodland.* Heidelberg: Springer.

Korte, M. and Mandea, M. (2008), "Magnetic Poles and Dipole Tilt Variation Over the Past Decades to Millennia", *Earth Planets Space*, 60, pp. 937–948.

Kosslyn, S. M. (1994), *Image and Brain: The Resolution of the Imagery Debate.* Cambridge, MA: MIT Press.

Kramer, G. (1959), "Recent Experiments on Bird Orientation", *Ibis*, 101, pp. 399–416.

Krause, E. (1893), *Die Trojaburgen Nordeuropas, ihr Zusammenhang mit der indogermanischen Trojasage von der entführten und gefangenen Sonnenfrau (Syrith, Brunhild, Ariadne, Helena), den Trojaspielen, Schwert- und Labyrinthtänzen zur Feier ihrer Lenzbefreiung.* Glogau: Carl Fleming.

Küster, H.-J. (1998), *Geschichte des Waldes. Von der Urzeit bis zur Gegenwart.* München: C. H. Beck.

Kvale, S. (1992), *Psychology and Postmodernism.* London: SAGE.

Kyselka, W. (1987), *An Ocean in Mind.* Honolulu: University of Hawaii Press.

La Barre, W. (1975), "Anthropological Perspectives on Hallucination and Hallucinogens", in Siegel, R. K. and West, L. J., *Hallucinations; Behavior, Experience, and Theory.* New York: John Wiley & Sons, pp. 9–52.

La Cecla, F. (2000), *Perdersi. L'uomo senza ambiente.* Bari: La Terza.

Lafitau, J. F. (1724), *Mœurs des sauvages ameriquains comparées aux mœurs des premiers temps.* Paris: Saugrain l'aîné.

Lahontan, L. A. de Lom d'Arce (1990 [1703]), *Oeuvres completes.* Montréal: Presse de L'Université de Montréal.

Lakoff, G. and Johnson, M. (2003 [1980]), *Metaphors We Live By.* Chicago: Chicago University Press.

Lane, F. C. (1963), "The Economic Meaning of the Invention of the Compass", *The American Historical Review*, 68, 3, pp. 605–617.

Lather, P. (2007), *Getting Lost. Feminist Efforts Towards a Double(d) Science.* Albany: State University of New York Press.

Lavalle, C. and Silva, F. (2011), Corine Land Cover 2006 - refined version. European Commission, Joint Research Centre (JRC) [Dataset] PID: http://data.europa.eu/89h/jrc-luisa-corine-land-cover-refined-2006. Accessed 10 April 2017.

Law, L. (1997), "Dancing on the Bar: Sex, Money, and the Uneasy Politics of Third Space," in Pile, S. and Keith, M. (Eds.), *Geographies of Resistance.* London: Routledge, pp. 124–151.

Lawton, C. A. (1994), "Gender Differences in Way-finding Strategies: Relationship to Spatial Ability and Spatial Anxiety", *Sex Roles*, 30, pp. 765–779.

Lawton, C. A. and Kallai, J. (2002), "Gender Differences in Wayfinding Strategies and Anxiety about Wayfinding: A Cross-cultural Comparison", *Sex Roles*, 47, 9–10, November 2002, pp. 389–401.

Lea, J. (2009), "Post-phenomenological Geographies", in Kitchen, R. and Thrift, N. (Eds.), *International Encyclopedia of Human Geography*. London: Elsevier.

Leach, N. (Ed.), *Rethinking Architecture: A Reader in Cultural Theory*. New York: Routledge, pp. 330–336.

Leão, L. (2000), "New Labyrinths and Maps: The Challenge of Cyberspace's Art", *Tenth International Symposium on Electronic Art (ISEA)*. Paris, Proceedings on line at http://isea2000.com.

Lefebvre, H. (1968), *La Vie quotidienne dans le monde modern*. Paris: Gallimard.

Lefebvre, H. (1970), *La Révolution urbaine*. Paris: Gallimard.

Lefebvre, H. (1980), *La presence et l'absence*. Paris: Castermann.

Lefebvre, H. (1991 [1974]), *The Production of Space* (Nicholson-Smith, D. Trans.). Oxford: Blackwell.

Lefebvre, H. (1995), *Writings on Cities* (Kofman, E. and Lebas, E. Eds.). Hoboken, NJ: Wiley-Blackwell.

Lefebvre, H. (2000 [1973]), *Le Droit à la ville II : Espace et politique*. Paris: Anthropos.

Lefebvre, H. (2004), *Rhythmanalysis: Space, Time and Everyday Life* (Elden, S. and Moore, G. Trans.). London: Continuum.

Lehmann, A. (1999), *Von Menschen und Baumen. Die Deutschen und ihr Wald*. Hamburg: Rotwohlt Verlag.

Lehmann, A. and Schriewer, K. (Eds.) (2000), *Der Wald – Ein deutscher Mythos? Perspektiven eines Kulturthemas*. Berlin: Dietrich Reimer Verlag.

Lejano, R., Ingram, M., and Ingram, H. (2013), *The Power of Narrative in Environmental Networks*. Cambridge, MA: MIT Press.

Levi, P. (1995 [1963]), *The Reawakening* (Woolf, S. Trans.). New York: Simon & Schuster.

Lewis, D. (1994 [1972]), *We, the Navigators: The Ancient Art of Landfinding in the Pacific*. Honolulu: University of Hawaii Press.

Lindow, J. (1994), "The Social Semantics of Cardinal Directions in Medieval Scandinavia", *The Mankind Quarterly*, 34, 3, Spring, pp. 209–224.

Lladò, B. (2013), *Franco Farinelli. Del mapa al laberinto*. Barcelona: Icaria. Espacios Criticos.

Lorimer, H. (2005), "Cultural Geography: The Busyness of Being 'More-than-Representational'", *Progress in Human Geography*, 29, 1, pp. 83–94.

Louden, R. B. (2011), "'The Play of Nature': Human Beings in Kant's Geography", in Elden, S. and Mendieta, E. (Eds.), *Reading Kant's Geography*. Albany: State University of New York Press, pp. 139–159.

Löw, M. (2001), *Raumsoziologie*. Frankfurt am Main: Suhrkamp Verlag.

Lubbock, J. (1882), *Ants, Bees, and Wasps: A Record of Observations on the Habits of the Social Hymenoptera*. London: Kegan Paul, & Trench.

Lubbock, J. (1913), *Prehistoric Times as Illustrated by Ancient Remains and the Manners and Customs of Modern Savages*. New York: Henry Holt.

Lund, F. H. (1930), "Physical Asymmetries and Disorientation", *The American Journal of Psychology*, 42, 1, pp. 51–62.

Luomala, K. (1958), "Review of Andrew Sharp, Ancient Voyagers in the Pacific", *American Anthropologist*, 60, pp. 776–778.

Luria, A. R. (1972), *The Man with a Shattered World: The History of a Brain Wound* (Foreword by Sacks, O.). Cambridge, MA: Harvard University Press.

Lussu, E. (1945), *Un anno sull'altipiano*. Turin: Einaudi.

Lynch, K. (1960), *The Image of the City*. Cambridge, MA: MIT Press.

Maccoby, E. E. and Jacklin, C. N. (1974), *The Psychology of Sex Differences*. Stanford: Stanford University Press.

Mach, E. (1897), "On Sensations of Orientation", *The Monist: A Quarterly Magazine Devoted to the Philosophy of Science*, Chicago, 8, pp. 79–96.

Mack, J. (2011), *The Sea: A Cultural History*. London: Reaktion Books.

Mackinder, H. J. (1902), *Britain and the British Seas*. New York: D. Appleton & Company.

Magnus, O. (1996 [1555]), *Description of the Northern Peoples* (Fisher, P. and Higgens, H. Trans.). London: Hakluyt Society.

Maguire, E. A., Woollet, K., and Spiers, H. G. (2006), "London Taxi Drivers and Bus Drivers: A Structural MRI and Neuropsychological Analysis", *HIPPOCAMPUS*, 16, pp. 1091–1101.

Maguire, E. A., Gadian, D. G., Johnsrude, I. S., Good, C. D., Ashburner, J., Frackowiak, R. S. J., and Frith, C. D. (2000), "Navigation-related Structural Change in the Hippocampi of Taxi Drivers", *Proc. Natl. Acad. Sci. USA*, 97, pp. 4398–4403.

Malpas, J. and Thiel, K. (2011), "Kant's Geography of Reason", in Elden, S. and Mendieta, E. (Eds.), *Reading Kant's Geography*. Albany: State University of New York Press, pp. 195–214.

Manovich, L. (2006), "The Poetics of Augmented Space", *Visual Communication*, 5, 2, pp. 219–240.

Mansel, P. (2010) *Levant: Splendour and Catastrophe on the Mediterranean*. London: John Murray.

Manzella, A. G. H. (2013), "Disorientation in Julie Otsuka's When the Emperor Was Divine. The Imprisoned Spaces of Japanese Americans during World War II", in Marouan, M. and Simmons, M. (Eds.), *Race and Displacement: Nation, Migration, and Identity in the Twenty-First Century*. Tuscaloosa, AL: University of Alabama Press, pp. 143–161.

Márai, S. (1998), *Embers* (Janeway, C. B. Trans.). New York: Vintage Books.

Marcuse, P. (1985), "Gentrification, Abandonment, and Displacement: Connections, Causes, and Policy Responses in New York City", *Journal of Urban and Contemporary Law*, 28, pp. 195–240. Available at: http://openscholarship.wustl.edu/law_urbanlaw/vol28/iss1/4. Accessed 9 April 2017.

Marcuzzi, M. (2011), "Writing Space. Historical Narrative and Geographical Description in Kant's Physical Geography", in Elden, S. and Mendieta, E. (Eds.), *Reading Kant's Geography*. Albany: State University of New York Press, pp. 115–136.

Marie, P. and Behague, P. (1919), "Syndrome de désorientation dans l'espace", *Revue Neurologique*, 26, 1, pp. 1–14.

Markula, P. and Pringle, R. (2006), *Foucault, Sport and Exercise: Power, Knowledge and Transforming the Self*. London: Routledge.

Marouan, M. and Simmons, M. (Eds.) (2013), *Race and Displacement: Nation, Migration, and Identity in the Twenty-First Century*, Toscaloosa: University of Alabama Press.

Marquardt, G. (2011), "Wayfinding for People With Dementia: A Review of the Role of Architectural Design", *HERD*, 4, 2, Winter, pp. 75–90.

Martin, N. and Rosello, M. (2016), "Disorientation: An Introduction", *Culture, Theory and Critique*, 57, 1, pp. 1–16.

Martinez, E. (2004) *The Tango Singer*. New York: Bloomsbury.

De Martino, E. (1977), *La fine del mondo. Contributo all'analisi delle apocalissi culturali*. Torino: Einaudi.

Masotti, A. (2012), "Dalla città labirinto alla città rizoma" *Orillas*, 1, Padua: Padova University Press.

Massey, D. (1994), *Space, Place and Gender*. London: Blackwell Publishing.

Massey, D. (2005), *For Space*. London: SAGE.

Massey, D., Allen, J., and Sarre, P. (Eds.) (1999), *Human Geography Today*. Cambridge: Polity Press.

Matthews, W. (1922), *Mazes and Labyrinths. A General Account of their History and Development*. London: Longmans, Green and Co.

Matvejevic, P. (1999), *Mediterranean. A Cultural Landscape*. Berkeley: University of California Press.

McBurney, D. H., Gaulin, S. J. C., Devineni, T., and Adams, C. (1997), "Superior Spatial Memory of Women: Stronger Evidence for the Gathering Hypothesis", *Evolutions and Human Behaviour*, 18, pp. 165–174.

McCarter, D. and Schmidt, R. (1998), *Lost! A Randger's Journal of Search and Rescue*. Yellow Springs, OH: Graphicom Press.

McDowell, L. (1992), "Doing Gender: Feminism and Research Methods in Human Geography", *Transactions of the Institute of British Geographers*, 17, 4, pp. 399–416.

McMains, J. (2006), *Glamour Addiction: Inside the American Ballroom Dance Industry*. Middletown: Wesleyan University Press.

McNamara, T. P., Rump, B., and Werner, S. (2003), "Egocentric and Geocentric Frames of Reference in Memory of Large-scale Space", *Psychon B Rev.*, 10, 3, pp. 589–595.

Medas S. (2004), *De Rebus Nauticis. L'arte della navigazione nel mondo antico*. Roma: L'Erma di Bretschneider.

Meloni, A. (1993), "La variazione secolare del campo geomagnetico", *Annali di Geofisica*, 36, Suppl. 5–6, pp. 41–54.

Melville, H. (1851), *Moby Dick*. New York: Bantam Books.

Mendieta, E. (2011), "Geography Is to History as Woman Is to Man: Kant on Sex, Race, and Geography", in Elden, S. and Mendieta, E. (Eds.), *Reading Kant's Geography*. Albany: State University of New York Press, pp. 345–368.

Menocal, M. R. (1987), *The Arabic Role in Medieval Literary History: A Forgotten Heritage*. Philadelphia: University of Pennsylvania Press.

Mentz, S. (2013), "God's Storms: Shipwreck and the Meanings of Ocean in Early Modern England and America", in Thompson, C. (Ed.), *Shipwreck in Art and Literature: Images and Interpretations from Antiquity to the Present Day*. New York, London: Routledge, pp. 77–91.

Merleau-Ponty, M. (1964), *L'oeil et l'Esprit*. Paris: Gallimard.

Merleau-Ponty, M. (2005 [1945]), *Phenomenology of Perception* (Smith, C. Trans.). London: Routledge.

Merleau-Ponty, M. (2006) [1964]), *L'oeil et l'Esprit*. Paris: Gallimard.

Meynem, E. (Ed.) (1985), *International Geographical Glossary*. Deutsche Ausgabe, Stuttgart: Frans Steiner Verlag.

Milner-Gulland, E. J., Fryxell, J. M., and Sinclair, A. R. E. (Eds.) (2011), *Animal Migration: A Synthesis*. Oxford: Oxford University Press.

Minaeva, O. and Holmquist, L. (Eds.) (2015), *Scandinavia and the Balkans. Cultural Interactions with Byzantium and Eastern Europe in the First Millennium AD*. Newcastle upon Tyne: Cambridge Scholars Publishing.

Minkowski, E. (1932) "Le problème de l'hallucination et le problème de l'espace (Quelques réflexions au sujet d'un halluciné)", *L'Évolution Psychiatrique*, 3, pp. 57–76.
Mitchell, W. J. (1996), *City of Bits: Space, Place, and the Infobahn*. Cambridge, MA: MIT Press.
Mitchell, W. J. (2013), *Iconology: Image, Text. Ideology*. Chicago: University of Chicago Press.
Moi, T. (1999), *What is a Woman and Other Essays*. Oxford: Oxford University Press.
Moita, M. A. P., Rosis, S., Zhou, Y., Le Doux, J. E., and Blair, H. T. (2004), "Putting Fear in Its Place: Remapping of Hippocampal Place Cells during Fear Conditioning", *The Journal of Neuroscience*, 24, 31, pp. 7015–7023.
Monmonier, M. (1991), *How to Lie with Maps*. Chicago: University of Chicago Press.
Montello, D. R., Lovelace, K. L., Golledge, R. G., and Self, C. M. (1999), "Sex-related Differences and Similarities in Geographic and Environmental Spatial Abilities", *Annals of the Association of American Geographers*, 89, 3, pp. 515–534.
Morris, D. (2004), *The Sense of Space*. Albany: State University of New York Press.
Morris, H. (1973), *Evolution and Modern Racism*, Impact Series No. 7, San Diego, California: Institute for Creation Research.
Muheim, R., Phillips, J. B., and Åkesson, S. (2006), "Polarized Light Cues Underlie Compass Calibration in Migratory Songbirds", *Science*, 313, pp. 837–839.
Müller-Sievers, H. (2003), *Desorientierung. Anatomie und Dichtung bei Georg Büchne*. Goettingen: Wallstein Verlag.
Munt, S. (1995), "The Lesbian Flâneur", in Bell, D. and Valentine, G., *Mapping Desire*. London: Routledge, pp. 114–125.
Murakami, H. (2005), *Kafka on the Shore*. London: Vintage.
Nash, C. (2000), "Performativity in Practice: Some Recent Work in Cultural Geography", *Progr. Hum. Geogr.*, 24, 653, pp. 653–664.
Neuwirth, R. (2005), *Shadow Cities. A Billion Squatters. A New Urban World*. New York: Routledge.
Newman, D. (2007), *An Overview of Spatial Disorientation as a Factor in Aviation Accidents and Incidents*, Aviation Research and Analysis Report. Canberra: Australian Government, Australian Transport Safety Bureau.
Nicholson, M. H. and Rousseau, G. S. (1968), *This Long Disease, My Life: Alexander Pope and the Sciences*. Princeton: Princeton University Press.
Njaradi, D. (2008), "Mevlevi Dervish Dance: 'Authentic' and/or 'transnational'", in Minors, H. J. and Morris, G. (Eds.), *Established Scholars Conference Proceedings*. London: Society for Dance Research, pp. 23–28.
Nuvolati, G. (2006), *Lo sguardo del vagabondo. Il flâneur e la città da Baudelaire ai postmoderni*. Bologna: Il Mulino.
Ó Nuallain, S. (Ed.), (2000), *Spatial Cognition: Foundations and Applications*. Amsterdam: John Benjamins.
O'Keefe, J. and Nadel, L. (1978), *The Hippocampus as a Cognitive Map*. Oxford: Oxford University Press.
O'Neill, O. (2011), "Orientation in Thinking: Geographical Problems, Political Solutions", in Elden, S. and Mendieta, E. (Eds.), *Reading Kant's Geography*. Albany: State University of New York Press, pp. 215–232.
Ohno, H. and Ohno Laboratory (Eds. and contribution) (2006), "Tokyo 2050 Fibercity", *The Japan Architect*, 63, Shinkenchiku-sha, Autumn.
Olsson, G. (2007), *Abysmal. A Critique of Cartographic Reason*. Chicago: Chicago University Press.

Orvieto, P. (2004), *Labirinti castelli giardini. Luoghi letterari di orrore e smarrimento.* Roma: Salerno.

Paci, M. (2002), *L'uomo e la foresta*. Roma: Meltemi Editore.

Pailhous, J. (1970), *La représentation de l'espace urbain. L'exemple du chauffeur de taxi*. Paris: P. U.F.

Paillard, J. (1991), *Brain and Space*. Oxford: Oxford Science Publications.

Pain, R. (1991), "Space, Sexual Violence and Social Control", *Progress in Human Geography*, 15, pp. 415–431.

Palmer, E. (1991), "Land of the Rising Sun: The Predominant East–West Axis among the Early Japanese", *Monumenta Nipponica*, 46, 1 pp. 69–90.

Passig, K. and Scholz, A. (2010), *Verirren. Eine Anleitung für Anfänger und Fortgeschrittene*. Berlin: Rowohlt.

Pease, A. and Pease, B. (1999), *Why Women Can't Read Maps … And Won't Stop Talking*. Mona Vale: Pease Training International.

Pease, A. and Pease, B. (2001), *Why Men Don't Listen and Women Can't Read Maps*. London: Orion.

Peraya, D. and Collard, A.-S. (2008), "La métaphore dans les campus virtuels. Des formes de la représentation à la 'représent-action'", *Recherches en Communication*, 29, pp. 57–80.

Phillips, R. (2000), *"Forest Beatniks" and "Urban Thoreaus": Gary Snyder, Jack Kerouac, Lew Welch, and Michael McClure*. Bern: Peter Lang Publishing.

Piaget, J. and Inhelder, B. (1997 [1948]), *The Child's Conception of Space*. Hove: Psychology Press.

Piccardi, L., Pizzamiglio, M. R., Bianchini, F., Palermo, L., Risetti, M., Zompanti, L., Guariglia, C., and D'Amico, S. (2011), *Orientamento Topografico nel Bambino*. Milano: Springer.

Plumwood, V. (1993), *Feminism and the Mastery of Nature*. London: Routledge.

Poincaré, H. (1895), "Analysis situs", *Journal de l'École Polytechnique*, 2, 1, pp. 1–123.

Porteous, A., (2002 [1928]), *The Forest in Folklore and Mythology*. Mineola: Dover Publications.

Postma, A., Kessel, R. P. C., and Van Asseln, M. (2004), "The Neuropsychology of Object-Location Memory", in Allen, G. L., *Human Spatial Memory: Remembering Where*. London: Lawrence Erlbaum Ass., pp. 143–160.

Pratolini, V. (1943), *Il Quartiere*. Firenze: Vallecchi.

Proust, M., (2006 [1927]), *Remembrance of Things Past*, (Trans. by C. K. Scott Moncrieff), Vol.1. London: Wordsworth Editions.

Puwar, N. (2004), *Space Invaders. Race, Gender and Bodies Out of Place*. Oxford: Berg.

Quadflieg, D. (2009), "Philosophie", in Günzel, S. (Ed.), *Raum-wissenschaften*. Frankfurt, pp. 274–289.

Queneau, R. (1959), *Zazie dans le métro*. Paris: Gallimard.

Raffestin, C. (1981 [1980]), *Pour une géographie du pouvoir*. Paris: Persée (Trad. It. *Per una geografia del potere*. Milano: Unicopli).

Raffestin, C. (1996), "Le Labyrinthe du monde", *Cahiers Vilfredo Pareto: Revue européenne des sciences sociales*, Tome 34, 104, pp. 111–124.

Raffestin, C. (undated), Mito e Riflessione scientifica. Ovvero il labirinto come metafora, typescript.

Recalcati, M. (2012), *Ritratti del desiderio*. Milan: Raffaello Cortina.

Reclus, E. (1894), "East and West", *Contemporary Review*, 66, 346, pp. 475–487.

Reed Doob, P. (1990), *The Idea of the Labyrinth from Classical Antiquity through the Middle Ages*. Ithaca: Cornell University Press.

Reisberg, B., Ferris, S. H., de Leon, M. J., and Crook, T. (1982), "The Global Deterioration Scale for Assessment of Primary Degenerative Dementia", *American Journal of Psychiatry*, 139, pp. 1136–1139.

Relph, E. (1970), "An Inquiry into the Relations between Phenomenology and Geography", *Canadian Geographer*, 14, pp. 193–201.

Relph, E. (1976), *Place and Placelessness*. London: Pion.

Resta, C. (1996), *Il luogo e le vie. Geografia del pensiero in Martin Heidegger*. Milano: Franco Angeli.

Resta, C. (1999), "Stranieri nella metropoli", *Anterem, Eterotopie*, 58, pp. 81–84.

Revill, G. (2004), "Cultural Geographies in Practice: Performing French Folk Music: Dance, Authenticity and Non-representational Theory", *Cultural Geographies*, 11, 199, pp. 199–209.

Ricoeur, P. (1975), *La métaphore vive*. Paris: Seuil.

Ricoeur, P. (2004), *The Rule of Metaphor: The Creation of Meaning in Language*. London: Routledge.

Rieser, J. J. (1999), "Dynamic Spatial Orientation and the Coupling of Representation and Action", in Golledge, R. (Ed.), *Wayfinding Behavior. Cognitive Mapping and Other Spatial Processes*. Baltimore: The John Hopkins University Press, pp. 168–190.

Rigaud, A. (2012), "Disorienting Geographies: Land Art and the American Myth of Discovery", *Miranda*, 6, Marking the Land in North America [on line], http://miranda.revues.org/2955.

Rorty, R. (1979), *Philosophy and the Mirror of Nature*. Princeton, NJ: Princeton University Press.

Rose, G. (1999), "Performing Space", in Massey, D., Allen, J., and Sarre, P. (Eds.), *Human Geography Today*. Cambridge: Polity Press, pp. 247–259.

Rosello, M. (2016), "Disorientation and Accompaniment: Paris, the Metro and the Migrant", *Culture, Theory and Critique*, 57, 1, pp. 77–91.

Rossi, P. (2008), "Daedalus sive mechanicus: humankind and machines", Convegno Internazionale Robotics, A New Science, Accademia Nazionale dei Lincei, 20 febbraiowww.castfvg.it/ZZ/roboetica/intervento_rossi_001.doc.

Rousseau, J.-J. (2012 [1754]), "Discourse on the Origin, and the Foundations of Inequality among Men", in Scott, J. T. (Trans. and Ed.), *The Major Political Writings of Jean-Jacques Rousseau: The Two "Discourses" and the "Social Contract"Discours sur l'origine et les fondements de l'inégalité parmi les homes*. Chicago: University of Chicago Press, pp. 37–152.

Rubenstein, N. (1993), "Environmental Autobiography and Places of Trauma", *Proceedings, 24th annual meeting, EDRA*. Oklahoma City: EDRA.

Ruiz-Gomez, M. Mollà (Ed.) (2010), *Lenguaje y visiones del paisaje y del territorio*. Madrid: UAM, Fundacion Duques de Soria.

Sachs, O. (2012), *Hallucinations*. London: Picador.

Said, E. W. (1978), *Orientalism*. New York: Pantheon.

Said, E. W. (1999), *Out of Place: A Memoir*. New York: Vintage Books.

De Saint-Hilaire, P. (1992), *L'universe Secret du Labyrinthe*. Paris: Robert Laffont.

Salazar, R. (2003), "The Labyrinth of the City. A Guided Visit to Make a Journey Around Different Interpretations of the City", *Ciencia Ergo Sum*, March, 10, 1, pp. 5–17.

Saldanha, A. (2005), "Trance and Visibility at Dawn: Racial Dynamics in Goa's Rave Scene", *Social & Cultural Geography*, 6, 5, October, pp. 707–721.

Sandercock, L. (1998), *Towards Cosmopolis: Planning for Multicultural Cities*. London: John Wiley.

Sandercock, L. (2003), *Cosmopolis II: Mongrel Cities in the 21st Century.* London: Continuum.

Santarcangeli, P. (1984), *Il libro dei labirinti.* Milano: Frassinelli.

Sassen, S. (1998), *Globalization and its discontents. Essays on the New Mobility of People and Money.* New York: New Press.

Sassen, S. (2001), *The Global City.* Princeton, NJ: Princeton University Press.

Saunders, T. (2016), "Metaphor and Disorientation", *Culture, Theory and Critique*, 57, 1, pp. 92–105.

De Saussure, L. (1923), *L'origine de la rose des vents et l'invention de la boussole.* Genève: Impr. Albert Kundig.

Saward, J. (2002), *Magical Paths: Labyrinths & Mazes in the 21st Century.* London: Mitchell.

Schaeffer, A. (1928), "Spiral Movement in Man", *Journal of Morphology and Physiology*, 45, 1, pp. 293–398.

Schama, S. (1995), *Landscape and Memory.* New York: Alfred A. Knopf.

Scharfe, M. (1998), *Wegzeiger. Zur Kulturgeschichte des Verirrens und Wegfindens.* Marburg: Jonas Verlag.

Schmidt di Friedberg, M. (2004), *L'arca di Noè. Conservazionismo tra natura e cultura.* Torino: Giappichelli.

Schmidt di Friedberg, M. (2007), "'Poco importa sapere per dove devi andare': il (dis) orientamento e le donne", *Geotema*, 33, pp. 27–33.

Schmidt di Friedberg, M. (2008), "La città. Un infinito limitato", *Scripta Nova*, Revista Electrònica de Geografìa y Ciencias Sociales, Universidad de Barcelona, 12, 270(61). www.ub.edu/geocrit/sn/sn-270/sn-270-61.htm. Accessed 9 April 2017.

Schmidt di Friedberg, M. (2010), "Mi ritrovai per una selva oscura chè la diritta via era smarrita. Pedersi nel bosco da Dante a Heidegger", in Cantero, N. Ortega, Alvarez, J. Garcìa, Ruiz-Gomez, M. Mollà (Eds.), *Lenguaje y visiones del paisaje y del territorio.* Madrid: UAM, Fundacion Duques de Soria, pp. 35–43.

Schmidt di Friedberg, M. (2012), "Perdersi nel cyberspazio. Il disorientamento virtuale", in Cerreti, C., Dumont, I., and Tabusi, M. (Eds.), *Geografia sociale. la sfida della comunicazione.* Roma: Aracne, pp. 121–131.

Schmidt di Friedberg, M. (2013), "Je ne suis pas là. La performance des lieux par la désorientation", in Kamal, M. and Raibaud, Y. (Eds.), *Genre et Construction de la Géographie.* Bordeaux: Maison des Sciences de l'Homme d'Acquitaine, pp. 27–38.

Schmidt di Friedberg, M. (2014), "Cet instinct qui sert à l'indigène de guide infaillible", in Alvarez, J. Garcia, Garcia, J. C. (Eds.), *Història da Geografia e Colonialismo.* Lisboa: Centro de Estudos Geogràficos da Universidade de Lisboa, pp. 77–88.

Schmidt di Friedberg, M. (2016) "Il labirinto del paesaggio", in de Pisòn, Martinez E. and Ortega, Cantero N. (Eds.), *El paisaje y sus simbolos.* Madrid: UAM, Fundacion Duques de Soria.

Schroer, M. (2006), *Räume, Orte, Grenzen. Auf dem Weg zu einer Soziologie des Raums.* Frankfurt am Mein: Suhrkamp.

Seamon, D. (1979), *A Geography of the Lifeworld.* New York: St. Martin.

Seton, M., Williams, S., and Zahirovic, S. (2013), "Obituary Sandy Island (1876–2012)", *Eos, Transactions, American Geophysical Union*, 94, 15, pp. 141–148.

Shannon, B. (2002), *The Antipodes of the Mind: Charting the Phenomenology of the Ayahuasca Experience.* Oxford: Oxford University Press.

Shapiro, G. (1995), *Earthwards: Robert Smithson and Art after Babel*. Berkeley: University of California Press.

Sharp, A. (1957), *Ancient Voyagers in the Pacific*. Harmondsworth: Penguin Books.

Sharp, A. (1963), *Ancient Voyagers in Polynesia*. Berkeley: University of California Press.

Shaw, P. (2008), *The (Mostly) True Story of Helvetica and the New York City Subway*; [online] AIGA VOICE. http://www.aiga.org/the-mostly-true-story-of-helvetica-and-the-new-york-city-subway. Accessed 2 April 2017.

Shusterman, R. (2008), *Body Consciousness: A Philosophy of Mindfulness and Somaesthetics*. New York: Cambridge University Press.

Simay, P. (2008), "Une autre ville pour une autre vie. Henri Lefebvre et les situationnistes", *Métropoles*, 4, pp. 202–213.

Simonsen, K. (2013), "In Quest of a New Humanism: Embodiment, Experience and Phenomenology as Critical Geography", *Progress in Human Geography*, 37(1), pp. 10–26.

Smith, M. (1998), "Modernism and the Avant-gardes", in Hill, J. and Church, G. P. (Eds.), *The Oxford Guide to Film Studies*. New York: Oxford University Press, pp. 395–410.

Smith, M., Davidson, J., Cameron, L., and Bondi, L. (2009), *Emotion, Place and Culture*. Aldershot: Ashgate.

Smith, Z. (2000), *White Teeth*. London: Hamish Hamilton.

Smithson, R. (1979), "A Museum of Language in the Vicinity of Art", in *The Writing of Robert Smithson* (Holt, N., Ed.). New York: New York University Press, pp. 74–76.

Smithson, R. (1996), *Robert Smithson: The Collected Writings* (Flam, J. D., Ed.). Oakland, CA: University of California Press.

Soja, E. (1996), *Thirdspace: Journeys to Los Angeles and Other Real-and-Imagined Places*. Oxford: Basil Blackwell.

Solnit, R. (2005), *A Field Guide to Getting Lost*. London: Viking.

Sonnenfeld, J. (1994), "Way-Keeping, Way-Finding, and Way-Losing: Disorientation in a Complex Environment", in Foote, K. E., Hugill, P. J., Mathewson, K., and Smith, J. M. (Eds.), *Re-reading Cultural Geography*. Austin: University of Texas Press, pp. 374–386.

Sonnenfeld, J. (2002), "Social Dimension of Geographic Disorientation in Artic Alaska", *Études/ Inuit/ Studies*, 26, 2, pp. 157–173.

Soper, K. (1995), *What is Nature? Culture, Politics and the Non-human*. Hoboken: Wiley & Sons.

Souman, J. L., Frissen, I., Sreenivasa, M. N., and Ernst, M. O. (2009), "Walking Straight into Circles", *Current Biology*, 19, 18, pp. 1538–1542.

Spencer, S. J., Steele, C. M., and Quinn, M. (1999), "Stereotype Threat and Women's Math Performance", *Journal of Experimental Social Psychology*, 35, pp. 4–28.

Spengler, O. (1991 [1918–1922]), *The Decline of the West* (Werner, H. Ed.). New York: Oxford University Press.

Spiers, H. J. and Maguire, E. A. (2006), "Thoughts, Behavior, and Brain Dynamics during Navigation in the Real World", *Neuroimage*, 31, pp. 1826–1840.

Squarcina, E. (2009), "Carta", in dell'Agnese, E., *Geo-grafia. Strumenti e parole*. Milano: Unicopli, pp. 85–94.

Stahel, D. (2009), *Operation Barbarossa and Germany's Defeat in the East*. Cambridge: Cambridge University Press.

Staszak, J-F. (2005), "Vos problèmes de couple expliqués par la géographie. Genre et espace dans quelques best-sellers", *Géographie et cultures*, 54, pp. 11–29.

Steele, C. M. and Aronson, J. (1995), "Stereotype Threat and the Intellectual Test Performance of African Americans", *Journal of Personality and Social Psychology*, 69, pp. 797–811.

Stegmaier, W. (Ed.) (2005), *Orientierung. Philosophische Perspektiven*. Berlin: de Gruyter.

Stegmaier, W. (2008), *Philosophie der Orientierung*. Berlin: de Gruyter.

Stern, E. and Portugali, J. (1999), "Environmental Cognition and Decision Making in Urban Navigation", in Golledge, R. (Ed.), *Wayfinding Behavior. Cognitive Mapping and Other Spatial Processes*. Baltimore: The John Hopkins University Press, pp. 99–120.

Steyerl, H. (2012), *The Wretched of the Screen*. Berlin: Stemberg Press.

Stiegler, B. (2009), *Technics and Time, 2: Disorientation*. Stanford: Stanford University Press.

Strahler, A. and Strahler, N. (1965), *Introduction to Physical Geography*. New York: John Wiley & Sons.

Straus, E. W. (1966), "The Upright Posture", *Phenomenological Psychology*. New York: Basic Books, pp. 137–165.

Stroppa, C. (2001), *Jan Amos Comenius e il sogno urbano*. Milano: Franco Angeli.

Sturluson, S. (2005), *The Prose Edda: Norse Mythology* (Byock, J. Trans.). London: Penguin.

Suarez, T. (2004), *Early Mapping of the Pacific. The Epic Story of Seafarers, Adventurers and Cartographers Who Mapped the Earth's Greatest Ocean*. North Clarendon, VT: Tuttle Publishing.

Svorou, S. (1993) *The Grammar of Space*. Amsterdam: John Benjamins.

Swift, J. (1900) [1726], *Gulliver's Travels into Several Remote Regions of the World*. Boston: D. C. Heath.

Swyngedouw, E. (1997), "Neither Global nor Local: 'Glocalization' and the Politics of Scale", in Cox, K. (Ed.), *Spaces of Globalization*. New York: Guilford, pp. 137–166.

Thein, D. (2005), "After or Beyond Feeling? A Consideration of Affect and Emotion in Geography", *Area*, 37, pp. 450–456.

Thevenin, R. and Coze, P. (1929), *Moeurs et histoire des indiens Peaux-Rouges*, 2 ed. Paris: Payot.

Thomas, J. R. and French, T. (1985), "Gender Differences Across Age in Motor Performance: A Meta-analysis", *Psychological Bulletin*, 98, pp. 260–282.

Thoreau, D. (1990 [1854]), *Walden*. Philadelphia: Courage Books.

Thrift, N. (1997), "The Still Point. Resistance, Expressive Embodiment and Dance", in Keith, M. and Pile, S. (Eds.), *Geographies of Resistance*. London: Routledge, pp. 124–151.

Thrift, N. (1999), "Steps Towards an Ecology of Place", in Massey, D., Allen, J., and Sarre, P. (Eds.), *Human Geography Today*. Cambridge: Polity Press, pp. 295–322.

Thrift, N. (2008), *Non-Representational Theory: Space, Politics, Affect*. London: Routledge.

Thrift, N. (2009), "Different Atmospheres: of Sloterdijk, China and Site", *Environment and Planning D: Society and Space*, 27, pp. 119–138.

Thrift, N. and Dewsbury, J.-D. (2000), "Dead Geographies – and How to Make Them Live", *Environment and Planning D: Society and Space*, 18, pp. 411–432.

Tolman, E. C. (1948), "Cognitive Maps in Rats and Men", *Psychological Review*, 55, pp. 189–208.

Totman, C. (1989), *The Green Archipelago: Forestry in Preindustrial Japan*. Berkley: University of California Press.

Du Tremolet de Lacheisserie, É., Gignoux, D., and Schlenker, M. (Eds.) (2003), *Magnetism: Fundamentals*. New York: Kluwer Academic Publishers.

Trowbridge, C. C. (1913), "On Fundamental Methods of Orientation and 'Imaginary Maps'", *Science*, 38, 990, pp. 888–896.

Tuan, Y.-F. (1974), *Topophilia. A Study Of Environmental Perception, Attitudes and Values*. Englewood Cliffs: Prentice-Hall.

Tuan, Y.-F. (2001), "The Desert and I: A Study in Affinity", *Michigan Quarterly Review*, 40, 1: Reimagining place, Winter..

Tuan, Y.-F. (2007), *Space and Place: The Perspective of Experience*. Minneapolis: University of Minnesota Press.

Tuan, Y.-F. (2009), *Who Am I: An Autobiography Of Emotion, Mind, And Spirit*. Madison: University of Wisconsin Press.

Twain, M. (1872), *Roughing It*. Hartford, CT: American Publishing Company.

Twain, M. (1880), *A Tramp Abroad*. London: Chatto & Windus.

United Nations Population Fund (2007), *State of the World Population 2007. Unleashing the Potential of Urban Growth*, UNFPA.

Upfield, A. (1956), *Man of Two Tribes*. London: Angus & Robertson.

Valentine, G. (1989), "The Geography of Women's Fear", *Area*, 21, pp. 385–390.

Valentine, G. (1996), "(Re)negotiating the 'Heterosexual Street': Lesbian Productions of Space", in Duncan, N. (Ed.), *Bodyspace: Destabilizing Geographies of Gender and Sexuality*. London: Routledge, pp. 145–155.

Van Cleve, J. and Frederick, R. E. (1991), *The Philosophy of Right and Left: Incongruent Counterparts and the Nature of Space*. Dordrecht: Springer.

Van Gennep, M. A. (1909), "Du sens d'orientation chez l'homme", *La Revue des Idées*, 70, pp. 290–301.

Veel, K. (2003), "The Irreducibility of Space: Labyrinths, Cities, Cyberspace, *Diacritics*, 33, 3/4, Fall-Fall-Winter, pp. 151–172.

Vicari, S. and Caselli, M. C. (Eds.) (2010), *Neuropsicologia dello sviluppo*. Bologna: Il Mulino.

Viguier, C. (1882), "Le sens de l'orientation et ses organes chez Les animaux et chez l'homme", *Revue Philosophique de la France Et de l'Etranger*, 14, pp. 1–36.

Virilio, P. (1995), "Vitesse Et Information. Alerte dans le cyberespace!", *Le Monde Diplomatique*, August.

Virilio, P. (1997 [1995]), *Open Sky* (Rose, J. Trans.). London: Verso.

Vogt, R. (1971), "Widerstandsformen in der CSSR als Antwort auf die Intervention der Warschauer-Pakt-Truppen vom 21. August 1968", *Soziale Verteidigung*, 3, 9/10, pp. 60–70.

Voyer, D., Nolan, C., and Voyer, S. (2000), "The Relation Between Experience And Spatial Performance in Men and Women – Statistical Data Included", *Sex Roles*, December, 43, 11–12, pp. 891–915.

Wallace, A. R. (1870), *Contributions to the Theory of Natural Selection. A Series of Essays*. London & New York: Macmillan & Co.

Waxman, W. (2014), *Kant's Anatomy of the Intelligent Mind*. Oxford: Oxford University Press.

Wehner, R. and Wehner, S. (1990), "Insect Navigation: Use of Maps or Ariadne's Thread?", *Ethol. Ecol. Evol.*, 2, pp. 27–48.

Werner, A. (1904), "Note on the Terms Used for 'Right Hand' and 'Left Hand' in the Bantu Languages", *Journal of the African Society*, 13, pp. 112–116.

Whatmore, S. J. (2002), *Hybrid Geographies*. London: SAGE.

Whorf, B. (1941), "The Relation of Habitual Thought and Behavior to Language", in Spier, L. (Ed.), *Language, Culture, and Personality, Essays in Memory of Edward Sapir.* Menasha, WI: Sapir Memorial Publication Fund, pp. 75–93.

Whorf, B. L. (1956), *Language, Thought, and Reality; Selected Writings.* Cambridge, MA: Technology Press of Massachusetts Institute of Technology.

Wilcove, D. S. (2008), *No Way Home: The Decline of the World's Great Animal Migrations.* Washington, DC: Island Press.

Wiltschko, W. and Wiltschko, R. (1988), "Magnetic Orientation in Birds", *Current Ornithology,* 5, pp. 67–121.

Winkler, O. (1899), "On Sea Charts Formerly Used in the Marshall Islands, with Notices on the Navigation of These Islanders in General," *Annual Report of the Board of Regents of the Smithsonian Institution*, 2 vols., Washington, DC: United States Government Printing Office, 1, pp. 487–509.

Withers, C. W. J. (2011), "Kant's Geography in Comparative Perspective", in Elden, S. and Mendieta, E. (Eds.), *Reading Kant's Geography.* Albany: State University of New York Press, pp. 47–65.

Wittgenstein, L. (1922), *Tractatus Logico-Philosophicus.* New York: Harcourt, Brace & Company.

Wittgenstein, L. (1958 [1953]), *Philosophical Investigations* (Anscombe, G. E. M. Trans.). Oxford: Basil Blackwell.

Wittgenstein, L. (1967), *Zettel* (Anscombe, G. E. M. and von Wright, G. H. Eds.). Berkeley: University of California Press.

Wrangell, F. (1841), *Narrative of an Expedition to the Polar Sea in the Years 1820, 1821 and 1823 commandes by lieutennt, now Admiral Ferdinand Wrangel of the Russian Imperial Navy.* New York: Harper and Brothers.

Wright, C. (2001), *The Maze and the Warrior.* Cambridge, MA: Harvard University Press.

Wurman, S. (1997), *Information Architects.* New York: Graphis.

Wyatt, G. W. (1876), *Myths and Songs from the South Pacific.* London: King.

Yardley, L. Ed. (2002), *Material Discourses of Health and Illness.* Hove: Psychology Press.

Young, I. M. (2005), *On Female Body Experience: "Throwing Like a Girl" and Other Essays.* Oxford: Oxford University Press.

Zambrano, M. (1988), *Claros del bosque.* Barcelona: Biblioteca del Bolsillo.

Zangl, V. (2011), "Remigration and Lost Time: Resuming Life after the Hocaust", in Creet, J. and Kitzmann, A. (Eds.), *Memory and Migration: Multidisciplinary Approaches to Memory Studies.* Toronto: University of Toronto Press, pp. 52–68.

Zinser, M. O., Palmer, D. L., and Miller, C. R. (2004), "Site Distance, Gender, and Knowledge of Geographic Sites", *Sex Roles,* 51, 11–12, December, pp. 661–686.

Index